MW00620217

The book of fallacies

From unfinished papers of Jeremy Bentham

Jeremy Bentham

A Friend

Alpha Editions

This edition published in 2019

ISBN : 9789353701871

Design and Setting By
Alpha Editions
email - alphaedis@gmail.com

This book is a reproduction of an important historical
work. Alpha Editions uses the best technology to
reproduce historical work in the same manner it was
first published to preserve its original nature. Any
marks or number seen are left intentionally to
preserve its true form.

THE

BOOK OF FALLACIES:

FROM

UNFINISHED PAPERS

OF

JEREMY BENTHAM.

BY A FRIEND.

LONDON:

PUBLISHED BY JOHN AND H. L. HUNT,

TAVISTOCK STREET, COVENT GARDEN.

—

1824.

LONDON:

PRINTED BY RICHARD TAYLOR, SHOE-LANE.

ALERE FLAMMAM

UNIV. OF CALIFORNIA
SOUTHERN BRANCH

JF
511
B44

EDITOR'S PREFACE.

THE substance of this treatise, drawn up from the most unfinished of all Mr. Bentham's Manuscripts, has already been published in French by M. Dumont; and considering the very extensive diffusion of that tongue, the present work, but for one consideration, might seem almost superfluous.

The original papers contain many applications of the writer's principles to British Institutions and British interests; which, with a view to continental circulation, have been judiciously omitted by M. Dumont.

To the English Reader, the matter thus omitted cannot but be highly important and

A 2

instructive. With the view of enabling him
to supply the deficiency, and to obtain separately a treatise of general importance, which
in the French work has somewhat unfortunately been appended to one of more limited
interest,—namely, that, on the mode of conducting business in Legislative Assemblies,
—the Editor has made the present attempt.

To have done justice to the original matter, the whole ought to have been re-written;
this, the Editor's other pursuits did not allow
him leisure to accomplish, and he has been
able to do little more than arrange the papers,
and strike out what was redundant.—In preparing the work for the press, Mr. Bentham
has had no share;—for whatever therefore
may be esteemed defective in the matter, or
objectionable in the manner, the Editor is
solely responsible. Still, he thought it better
that the work should appear, even in its present shape, than not appear at all; and having
devoted to it such portion of his time as could

be spared from the intervals of a life of labour, he hopes he shall not be without acknowledgement, from those who are competent to appreciate the value of whatsoever comes from the great founder of the science of morals and legislation.

M. Dumont's work contains an examination of the declaration of the Rights of Man, as proclaimed by the French Constituent Assembly. This forms no part of the present volume, to the subject of which, indeed,—Fallacies employed in *debate*,—it is not strictly pertinent. But, in fact, the original papers have been mislaid, and they seemed to lose so much of their spirit in a translation from the French, that the contents of the additional chapter would not compensate for the additional bulk and expense of the book.

CONTENTS.

INTRODUCTION.

PART THE FIRST.

FALLACIES OF AUTHORITY,

The subject of which is Authority in various shapes; and the object, to repress all exercise of the reasoning Faculty.

PART THE SECOND.

FALLACIES OF DANGER,

The subject matter of which is Danger in various shapes; and the object, to repress Discussion altogether, by exciting alarm.

CONTENTS.

INTRODUCTION.

Section I.

A FALLACY, WHAT.

BY the name of *fallacy* it is common to designate any argument employed, or topic suggested, for the purpose, or with a probability, of producing the effect of deception,—of causing some erroneous opinion to be entertained by any person to whose mind such argument may have been presented.

Section II.

FALLACIES, BY WHOM TREATED OF HERETOFORE.

The earliest author extant in whose works any mention is made on the subject of *fallacies* is Aristotle; by whom, in the course or rather on the occasion of his treatise on Logic, not only is this subject started, but a list of the species of arguments to which this denomination is applicable is undertaken to be given. Upon the principle of the exhaustive method at so early a period employed by that astonishing genius, and in comparison of what it might and ought to have been, so little turned to account since, *two* is the

B

number of parts into which the whole mass is distributed—fallacies in the diction, fallacies not in the diction : and thirteen (whereof in the diction six, not in the diction seven) is the number of the articles distributed between those two parts [a].

As from Aristotle down to Locke, on the subject of the origination of our ideas, (deceptious and undeceptious included,)—so from Aristotle down to this present day, on the subject of the forms, of which such ideas or combination of ideas as are employable in the character of instruments of deception, are susceptible,—all is a blank.

To do something in the way of filling up this blank is the object of the present work.

In speaking of Aristotle's collection of fallacies, as a stock to which from his time to the present no addition has been made, all that is meant, is, that whatsoever arguments may have had deception for their object, none besides those brought to view by Aristotle, have been brought to view in that character and under that name; for between the time of Aristotle and the present, treatises of the art of oratory, or popular argumentation, have not been wanting in various

---

[a] Σοφισμα, from whence our English word *sophism*, is the word employed by Aristotle. The choice of the appellation is singular enough ; σοφος is the wo d that was already in use for designating a wise man. It was the same appellation that was commonly employed for the designation of the seven Sages. Σοφιστης, from whence our *sophist*, being an impretative of Σοφος, was the word applied as it were in irony to designate the tribe of wranglers, whose pretension to the praise of wisdom had no better ground than an abuse of words.

languages and in considerable number : nor can any of these be found in which, by him who may wish to put a deceit upon those to whom he has to address himself, instruction in no small quantity may not be obtained.

What in these books of instruction is professed to be taught comes under this general description : viz. how, by means of words aptly employed, to gain your point ; to produce upon those with whom you have to deal, those to whom you have to address yourself, the impression, and, by means of the impression, the disposition most favourable to your purpose, whatsoever that purpose may be.

As to the impression and disposition the production of which might happen to be desired, whether the impression were correct or deceptious, whether the disposition were with a view to the individual or community in question, salutary, indifferent, or pernicious, was a question that seemed not in any of these instances to have come across the author's mind. In the view taken by them of the subject, had any such question presented itself, it would have been put aside as foreign to the subject ; exactly as, in a treatise on the art of war, a question concerning the justice of the war.

Dionysius of Halicarnassus, Cicero, and Quintilian, Isaac Voss, and, though last and in bulk least, yet not the least interesting, our own Gerard Hamilton (of whom more will be said), are of this stamp.

Between those earliest and these latest of the writers

who have written on this subject and with this view, others in abundance might be inserted ; but these are quite enough.

After so many ages past in teaching with equal complacency and indifference the art of true instruction and the art of deception—the art of producing good effects, and the art of producing bad effects—the art of the honest man and the art of the knave—of promoting the purposes of the benefactor and the purposes of the enemy of the human race ;—after so many ages during which, with a view to persuasion, disposition, action, no instructions have been endeavoured to be given but in the same strain of imperturbable impartiality, it seemed not too early in the nineteenth century to take up the subject on the ground of morality, and to invite common honesty for the first time to mount the bench and take her seat as judge.

As to Aristotle's fallacies, unless his *petitio principii* and his *fallacia, non causa pro causâ* be considered as exceptions, upon examination so little danger would be found in them, that, had the philosopher left them unexposed to do their worst, the omission need not have hung very heavy upon his conscience ; scarce in any instance will be discovered any the least danger of final deception : the utmost inconvenience they seem capable of producing seems confined to a slight sensation of embarrassment. And as to the embarrassment, the difficulty will be, not in pronouncing that the proposition in question is incapable of forming a just ground for the conclusion built upon it, but

in finding words for the description of the weakness
which is the cause of this incapacity : not in discover-
ing the proposition to be absurd, but in giving an
exact description of the form in which the absurdity
presents itself.

Section III.

RELATION OF FALLACIES TO VULGAR ERRORS.

Error, vulgar error [a], is an appellation given to an
opinion which, being considered as false, is considered
in itself only, and not with a view to any consequences
of any kind, of which it may be productive.

It is termed *vulgar* with reference to the persons
by whom it is supposed to be entertained : and this
either in respect of their multitude, simply, or in respect of the lowness of the station occupied by them
or the greater part of them in the scale of respectabi-
lity, in the scale of intelligence.

Fallacy is an appellation applied not exclusively to
an opinion or to propositions enunciative of supposed
opinions, but to discourse in any shape considered as
having a tendency, with or without design, to cause
any erroneous opinion to be embraced, or even, through

---

[a] *Vulgar errors* is a denomination which, from the work written on
this subject by a physician of name in the seventeenth century, has
obtained a certain degree of celebrity.

Not the moral (of which the political is a department), but the physical, was the field of the errors which it was the object of Sir Thomas
Brown to hunt out and bring to view : but of this restriction no intimation is given by the words of which the title of his work is composed.

the medium of erroneous opinion already entertained,
to cause any pernicious course of action to be engaged
or persevered in.

Thus, to believe that they who lived in early or old
times were, because they lived in those times, wiser
or better than those who live in later or modern times,
is vulgar error : the employing that vulgar error in
the endeavour to cause pernicious practices and institutions to be retained, is fallacy.

By those by whom the term *fallacy* has been employed, at any rate by those by whom it was originally
employed, deception has been considered not merely
as a consequence more or less probable, but as a consequence the production of which was aimed at on
the part at least of some of the utterers.

Ελεγχοι σοφισων, arguments employed by the sophists,
is the denomination by which Aristotle has designated
his devices, thirteen in number, to which his commen-
tators, such of them as write in Latin, give the name
of *fallaciæ*, (from *fallere* to deceive,) from which our
English word *fallacies*.

That in the use of these instruments, such a thing
as deception was the object of the set of men mentioned by Aristotle under the name of sophists, is altogether out of doubt. On every occasion on which
they are mentioned by him, this intention of deceiving
is either directly asserted or assumed.

Section IV.

POLITICAL FALLACIES THE SUBJECT OF THIS WORK.

The present work confines itself to the examination and exposure of only one class of fallacies, which class is determined by the nature of the occasion in which they are employed.

The occasion here in question is that of the formation of a decision procuring the adoption or rejection of some measure of *government :* including under the notion of a measure of government, a measure of legislation as well as of administration ; two operations so intimately connected, that the drawing of a boundary line between them will in some instances be matter of no small difficulty, but for the distinguishing of which on the present occasion, and for the purpose of the present work, there will not be any need.

Under the name of a *Treatise on Political Fallacies*, this work will possess the character, and, in so far as the character answers the design of it, have the effect of a treatise on *the art of government :* having for its practical object and tendency, in the first place, the facilitating the introduction of such features of good government as remain to be introduced ; in the next place giving them perpetuation—perpetuation, not by means of legislative clauses aiming directly at that object (an aim of which the inutility and mischievousness will come to be fully laid open to view in the course of this work), but by means of that instrument, viz.

reason, by which alone the endeavour can be productive of any useful effect.

Employed in this endeavour, there are two ways in which this instrument may be applied : one, the more direct, by showing, on the occasion of each proposed measure, in what way, by what probable consequences it tends to promote the accomplishment of the end or object which it professes to have particularly in view : the other, the less direct, by pointing out the irrelevancy, and thus anticipating and destroying the persuasive force, of such deceptious arguments as have been in use, or appear likely to be employed in the endeavour to oppose it, and to dissuade men from concurring in the establishment of it.

Of these two different but harmonizing modes of applying this same instrument to its several purposes, the *more direct* is that of which a sample has, ever since the year 1802, been before the public, in that collection of unfinished papers on legislation, published at *Paris* in the French language, and which had the advantage of passing through the hands of Mr. Dumont, but for whose labours it would scarcely, in the author's life-time at least, have seen the light. To exhibit the *less direct*, but in its application the more extensive mode, is the business of the present work.

To give existence to good arguments was the object in that instance : to provide for the exposure of bad ones is the object in the present instance—to provide for the exposure of their real nature, and thence for the destruction of their pernicious force.

Sophistry is a hydra of which, if all the necks could be exposed, the force would be destroyed. In this work they have been diligently looked out for, and in the course of it the principal and most active of them have been brought to view.

Section V.

DIVISION OR CLASSIFICATION OF FALLACIES.

So numerous are the instruments of persuasion which in the character of fallacies the present work will bring to view, that, for enabling the mind to obtain any tolerably satisfactory command over it, a set of divisions deduced from some source or other appeared to be altogether indispensable.

To frame these divisions with perfect logical accuracy will be an undertaking of no small difficulty ; an undertaking requiring more time than either the author or editor has been able to bestow upon it.

An imperfect classification, however, being preferable to no classification at all, the author had adopted one principle of division from the situation of the utterers of fallacies, especially from the utterers in the British Houses of Parliament :—fallacies of the *ins*,—fallacies of the *outs*,—*either-side* fallacies.

A principle of subdivision he found in the quarter to which the fallacy in question applied itself, in the persons on whom it was designed to operate; the *affections*, the *judgement*, and the *imagination*.

To the several clusters of fallacies marked out by this subdivision, a Latin affix, expressive of the faculty

or affection aimed at, was given ; not surely for osten-
tation, for of the very humblest sort would such osten-
tation be, but for *prominence,* for impressiveness, and
thence for clearness : arguments 1. *ad verecundiam :*
2. *ad superstitionem:* 3. *ad amicitiam:* 4. *ad metum :*
5. *ad odium :* 6. *ad invidentiam :* 7. *ad quietem :*
8. *ad socordiam :* 9. *ad superbiam :* 10. *ad judicium :*
11. *ad imaginationem.*

In the same manner, *Locke* has employed Latin denominations to distinguish four kinds of argument : *ad
verecundiam, ad ignorantiam, ad hominem, ad judicium.*

Mr. Dumont, who some few years since published
in French a translation, or rather a *redaction,* of a considerable portion of the present work, divided the fallacies into three classes, according to the particular
or special object to which the fallacies of each class
appeared more immediately applicable. Some he
supposed destined to repress discussion altogether ;
others to postpone it ; others to perplex, when discussion could no longer be avoided. The first class he
called fallacies of *authority,* the second fallacies of
delay, and the third fallacies of *confusion :* he has
also added to the name of each fallacy the Latin affix
which points out the faculty or affection to which it
is chiefly addressed.

The present editor has preferred this arrangement
to that pursued by the author, and with some little
variation he has adopted it in this volume.

In addition to the supposed immediate *object* of a
given class of fallacies, he has considered the *subject*

matter of each individual fallacy, with a view to the comprehending in one class all such fallacies as more nearly resemble each other in the nature of their subject matter : and the classes he has arranged in the order in which the enemies of improvement may be supposed to resort to them according to the emergency of the moment.

First, fallacies of *authority* (including laudatory personalities) ; the subject matter of which is authority in various shapes, and the immediate object, *to repress*, on the ground of the weight of such authority, *all exercise of the reasoning faculty.*

Secondly, fallacies of *danger* (including vituperative personalities) ; the subject matter of which is the suggestion of danger in various shapes, and the object, to *repress altogether*, on the ground of such danger, the *discussion* proposed to be entered on.

Thirdly, fallacies of *delay;* the subject matter of which is an assigning of reasons for delay in various shapes, and the object, to *postpone* such *discussion*, with a view of eluding it altogether.

Fourthly, fallacies of *confusion;* the subject matter of which consists chiefly of vague and indefinite generalities, while the object is *to produce*, when discussion can no longer be avoided, such *confusion* in the minds of the hearers as to incapacitate them for forming a correct judgement on the question proposed for deliberation.

In the arrangement thus made, imperfections will be found, the removal of which, should the removal of them be practicable and at the same time worth the trouble, must be left to some experter hand. The classes themselves are not in every instance sufficiently distinct from each other; the articles ranged under them respectively not appertaining with a degree of propriety sufficiently exclusive to the heads under which they are placed. Still, imperfect as it is, the arrangement will, it is hoped, be found by the reflecting reader not altogether without its use.

Section VI.

NOMENCLATURE OF POLITICAL FALLACIES.

Between the business of classification and that of nomenclature, the connexion is most intimate. To the work of classification no expression can be given but by means of nomenclature: no name other than what in the language of grammarians is called a *proper* name, no name more extensive in its application than is the name of an individual, can be applied, but a class is marked out, and, as far as the work of the mind is creation, *created.*

Still, however, the two operations remain not the less distinguishable: for of the class marked out, a description may be given of any length and degree of complication; the description given may be such as to occupy entire sentences in any number. But a name properly so called consists either of no more than one word, and that one a noun substantive, or at most of no more than a substantive with its adjunct:

or, if of words more than one, they must be in such sort linked together as to form in conjunction no more than a sort of compound word, occupying the place of a noun substantive in the composition of a sentence.

Without prodigious circumlocution and inconvenience, a class of objects, however well marked out by description, cannot be designated, unless we substitute for the words constituting the description, a word, or very small cluster of words, so connected as to constitute a name. In this case nomenclature is to description what, in algebraical operation, the substitution of a single letter of the alphabet for a line of any length composed of numerical figures or letters of the alphabet, or both together, is to the continuing and repeating at each step the complicated matter of that same line.

The class being marked out whether by description or denomination, an operation that will remain to be performed is, if no name be as yet given to it, the finding for it and giving to it a name : if a name has been given to it, the sitting in judgement on such name, for the purpose of determining whether it presents as adequate a conception of the object as can be wished, or whether some other may not be devised by which that conception may be presented in a manner more adequate.

Blessed be he for evermore, in whatsoever robe arrayed, to whose creative genius we are indebted for the first conception of those too short-lived vehicles, by which, as in a nutshell, intimation is conveyed to us of the essential character of those awful volumes which at the touch of the sceptre become the rules of our conduct, and the arbiters of our destiny :—" The Alien Act," "The Turnpike Act," "The Middlesex Waterworks Bill," &c. &c. !

How advantageous a substitute in some cases, how useful an additament in all cases, would they not make to those authoritative masses of words called *titles*, by which so large a proportion of sound and so small a proportion of instruction are at so large an expense of attention granted to us ; " An Act to explain and amend an Act entitled An Act to explain and amend," &c. &c. !

In two, three, four, or at the outside half a dozen words, information without pretension is given, which frequently when pretended is not given, but confusion and darkness given instead of it, in twice, thrice, four times, or half a dozen times as many lines.

Rouleaus of commodious and significative appellatives, are thus issued day by day throughout the session from an invisible though not an unlicensed mint; but no sooner has the last newspaper that appeared the last day of the session made its way to the most distant of its stages, than all this learning, all this circulating medium, is as completely lost to the world and buried in oblivion as a French assignat.

So many yearly strings of words, not one of which is to be found in the works of Dryden, with whom the art of coining words fit to be used became numbered among the lost arts, and the art of giving birth to new ideas among the prohibited ones ! So many words, not one of which would have found toleration from the orthodoxy of Charles Fox !

Let the workshop of invention be shut up for ever, rather than that the tympanum of taste should be grated by a new sound ! Rigorous decree ! more rigorous if obedience or execution kept pace with design, than even the continent-blockading and commerce-crushing decrees proclaimed by Buonaparte.

So necessary is it that, when a thing is talked of, there should be a name to call it by ; so conducive, not to say necessary, to the prevalence of reason, of common sense, and moral honesty, that instruments of deception should be talked of, and well talked of, and talked out of fashion,—in a word talked down,— that, without any other license than the old one granted by Horace, and which, notwithstanding the acknowledged goodness of the authority, men are so strangely backward to make use of,—the author had, under the spur of necessity, struck out for each of these instruments of deception a separate barbarism, such as the tools which he had at command would enable him to produce : the objections, however, of a class of readers, who, under the denomination of *men of taste*, attach much more importance to the manner than to the matter of a composition, have induced the editor to suppress for the present some of these characteristic appellations, and to substitute for them a less expressive periphrasis.

Section VII.

CONTRAST BETWEEN THE PRESENT WORK AND HAMILTON'S "PARLIAMENTARY LOGIC."

Of this work, the general conception had been formed, and in the composition of it some little progress made, when the advertisements brought under the author's notice the posthumous work intituled "*Parliamentary Logic*, by the late William Gerard Hamilton," distinguished from so many other Hamiltons by the name of *Single-speech Hamilton*.

Of finding the need of a work such as the present, superseded in any considerable degree by that of the right honourable orator, the author had neither hope nor apprehension: but his surprise was not inconsiderable on finding scarcely in any part of the two works any the smallest degree of coincidence.

In respect of practical views and objects, it would not indeed be true to say that between the one and the other there exists not any relation; for there exists a pretty close one, namely, the relation of contrariety.

When, under the title of "*Directions to Servants*," Swift presented to view a collection of such various faults as servants, of different descriptions, had been found or supposed by him liable to fall into, his object (it need scarce be said), if he had any serious object beyond that of making his readers laugh, was, not that compliance, but that non-compliance, with the directions so humorously delivered, should be the practical result.

Taking that work of Swift's for his pattern, and what seemed the serious object of it, for his guidance, the author of this work occasionally found in the form of a direction for the framing of a fallacy, what seemed the most convenient vehicle for conveying a conception of its nature : as in some instances, for conveying a conception of the nature of the figure he is occupied in the description of, a mathematician begins with giving an indication of the mode in which it may be framed, or, as the phrase is, *generated.*

On these occasions much pains will not be necessary to satisfy the reader that the object of any instructions which may here be found for the composition of a fallacy, has been, not to promote, but as far as possible to prevent the use of it: to prevent the use of it, or at any rate to deprive it of its effect.

Such, if Gerard Hamilton is to be believed, was not the object with Gerard Hamilton : his book is a sort of school, in which the means of advocating what is a good cause, and the means of advocating what is a bad cause, are brought to view with equal frankness, and inculcated with equal solicitude for success : in a word, that which Machiavel has been supposed sometimes to aim at, Gerard Hamilton as often as it occurs to him does not only aim at, but aim at without disguise. Whether on this observation any such imputation as that of calumny is justly chargeable, the samples given in the course of this work will put the reader in a condition to judge.

Sketched out by himself and finished by his editor and panegyrist[a], the political character of Gerard Hamilton may be comprised in a few words : he was determined to join with a party; he was as ready to side with one party as another; and whatever party he sided with, as ready to say any one thing as any other in support of it. Independently of party, and personal profit to be made from party, right and wrong, good and evil, were in his eyes matters of indifference. But having consecrated himself to party, viz. the party, whatever it was, from which the most was to be got, —that party being, of whatever materials composed, the party of the *ins*,—that party standing constantly pledged for the protection of abuse in every shape,

---

[a] Extract from the preface to Hamilton's work :—

"He indeed considered politics as a kind of game, of which the stake or prize was the administration of the country. Hence he thought that those who conceived that one party were possessed of greater ability than their opponents, and were therefore fitter to fill the first offices in the state, might with great propriety adopt such measures (consistent with the Constitution) as should tend to bring their friends into the administration of affairs, or to support them when invested with such power, without weighing in golden scales the particular parliamentary questions which should be brought forward for this purpose; as on the other hand, they who had formed a higher estimate of the opposite party might with equal propriety adopt a similar conduct, and shape various questions for the purpose of showing the imbecility of those in power, and substituting an abler ministry, or one that they considered abler, in their room ; looking on such occasions rather to the object of each motion than to the question itself. And in support of these positions, which, however short they may be of theoretical perfection, do

and in so far as good consists in the *extirpation* of abuse, for the opposing and keeping out every thing that is good,—hence it was to the opposing of whatsoever is good in honest eyes, that his powers, such as they were, were bent and pushed with peculiar energy.

One thing only he recognised as being *malum in se*, as a thing being to be opposed at any rate, and at any price, even on any such extraordinary supposition as that of its being brought forward by the party with which, at the time being, it was his lot to side. This was, *parliamentary reform*.

In the course of his forty years labour in the service of the people, one thing he did that was good : one thing to wit, that in the account of his panegyrist is set down on that side:—

---

not perhaps very widely differ (says Mr. Malone) from the actual state of things, he used to observe that, if any one would carefully examine all the questions which have been agitated in Parliament from the time of the Revolution, he would be surprised to find how *few* could be pointed out in which an honest man might not conscientiously have voted on either side, however, by the force of rhetorical aggravation and the fervour of the times, they may have been represented to be of such high importance, that the very existence of the state depended on the result of the deliberation.

" Some questions, indeed, he acknowledged to be of a vital nature, of such magnitude, and so intimately connected with the safety and welfare of the whole community, that no inducement or friendly disposition to any party ought to have the smallest weight in the decision. One of these in his opinion was the proposition for a *parliamentary reform*, or in other words for the new modelling the constitution of parliament; a measure which he considered of such moment, and of so dangerous a tendency, that he once said to a friend now living, that he would sooner suffer his right hand to be cut off than vote for it."

One use of government (in eyes such as his the principal use) is to enable men who have shares in it to employ public money in payment for private service:—

Within the view of Gerard Hamilton there lived a man whose talents and turn of mind qualified him for appearing with peculiar success in the character of an amusing companion in every good house. In this character he for a length of time appeared in the house of Gerard Hamilton: finding him an Irishman, Hamilton got an Irish pension of 300l. a year created for him, and sent him back to Ireland: the man being in Dublin, and constituting in virtue of his office a part of the lord lieutenant's family, he appeared in the same character and with equal success in the house of the lord lieutenant [a].

---

[a] "Yet, such was the warmth of his friend's feelings, and with such constant pleasure did he reflect on the many happy days which they had spent together, that he not only in the first place obtained for him a permanent provision on the establishment of Ireland *, but in addition to this proof of his regard and esteem, he never ceased, without any kind of solicitation, to watch over his interest with the most lively solicitude; constantly applying in person on his behalf to every new lord lieutenant, if he were acquainted with him; or if that were not the case, contriving by some circuitous means to procure Mr. Jephson's re-appointment to the office originally conferred on him by Lord Townshend: and by these means chiefly he was continued for a long series of years under twelve successive governors of Ireland in the same station, which had always before been considered a temporary office."— *Parl. Log.* 44.

---

* *Note by editor Malone :*—" A pension of 300l. a year, which the Duke of Rutland during his government, from personal regard and a high admiration of Mr. Jephson's talents, increased to 600l. *per annum* for the joint lives of himself and Mrs. Jephson. He survived our author

His Grace gave permanence to the sinecure, and doubled the salary of it. Here was liberality upon liberality—here was virtue upon virtue. It is by such things that merit is displayed; it is for such things that taxes are imposed; it is for affording matter and exercise for such virtues; it is for affording rewards for such merit, that the people of every country, in so far as any good use is made of them, are made.

To a man in whose eyes public virtue appeared in this only shape, no wonder that parliamentary reform should be odious:—of parliamentary reform, the effect of which, and, in eyes of a different complexion, one main use would be, the drying up the source of all such virtues.

Here, in regard to the matter of fact, there are two representations given of the same subject: representations perfectly concurrent in all points with one another, though from very different quarters, and beginning as well as ending with very different views, and leading to opposite conclusions.

Parliament a sort of gaming-house; members on the two sides of each house the players; the property of the people, such portion of it as on any pretence may be found capable of being extracted from them, the stakes played for. Insincerity in all its shapes, disingenuousness, lying, hypocrisy, fallacy, the instru-

---

but a few years, dying at his house at Black Rock, near Dublin, of a paralytic disorder, May 31, 1803, in his sixty-seventh year."

Note.—That not content with editing, and, in this way, recommending in the lump these principles of his friend and countryman, Malone takes up particular aphorisms, and applies his mind to the elucidation of them. This may be seen exemplified in Aphorisms 243, 249.

ments employed by the players on both sides for obtaining advantages in the game : on each occasion,—in
respect of the side on which he ranks himself,—what
course will be most for the advantage of the universal
interest,—a question never looked at, never taken into
account : on which side is the prospect of personal advantage in its several shapes,—this the only question
really taken into consideration : according to the answer given to this question in his own mind, a man
takes the one or the other of the two sides : the side
of those in office, if there be room or near prospect
of room for him ; the side of those by whom office
is but in expectancy, if the future contingent presents
a more encouraging prospect than the immediately
present.

To all these distinguished persons, to the self-ap-
pointed professor and teacher of political profligacy,
to his admiring editor, to their common and sympa-
thizing friend [a], the bigotry-ridden preacher of hollow
and common-place morality, parliamentary reform we
see in an equal degree, and that an extreme one, an
object of abhorrence.   How should it be otherwise ?
By parliamentary reform, the prey, the perpetually re-
nascent prey, the fruit and object of the game, would
have been snatched out of their hands.   Official pay
in no case more than what is sufficient for the security
of adequate service,—no sinecures, no pensions, for
hiring flatterers and pampering parasites :—no plunder-
ing in any shape or for any purpose :—amidst the cries

---

[a] See post 26.

of No theory! no theory! the example of America a
lesson, the practice of America transferred to Britain.

The notion of the general predominance of self-regarding over social interest has been held up as a
weakness incident to the situation of those whose converse has been more with books than men. Be it so :
look then to those teachers, those men of practical
wisdom, whose converse has been with men at least
as much as with books : look in particular to this
right honourable, who in the house of commons had
doubled the twenty years lucubration necessary for
law, who had served almost six apprenticeships, who
in that office had served out five complete clerkships:
what says he ? Self-regarding interest predominant
over social interest ? Self regard predominant ? no :
but self-regard sole occupant : the universal interest,
howsoever talked of, never so much as thought of;
right and wrong, objects of avowed indifference.

Of the self-written Memoirs of Bubb Dodington
how much was said in their day ! of Gerard Hamilton's Parliamentary Logic, how little ! The reason is
not unobvious: Dodington was all anecdote; Hamilton was all theory. What Hamilton endeavoured to
teach with Malone and Johnson for his bag-bearers,
Dodington was seen to practise.

Nor is the veil of decorum cast off any where from
his practice. In Hamilton's book for the first time
has profligacy been seen stark naked. In the reign
of Charles the Second, Sir Charles Sedley and others
were indicted for exposing themselves in a balcony in

a state of perfect nudity. In Gerard Hamilton may be seen the Sir Charles Sedley of political morality. Sedley might have stood in his balcony till he was frozen, and nobody the better, nobody much the worse: but Hamilton's self-exposure is most instructive.

Of parliamentary reform were a man to say that it is good because Gerard Hamilton was averse to it, he would fall into the use of one of those fallacies against the influence of which it is one of the objects of the ensuing work to raise a barrier;

This however may be said, and said without fallacy, viz. that it is the influence exercised by such men, and the use to which such their influence is put by them, that constitutes no small part of the political disease, which has produced the demand for parliamentary reform in the character of a remedy.

To such men it is as natural and necessary that parliamentary reform should be odious, as that Botany Bay or the Hulks should be odious to thieves and robbers.

Above all other species of business, the one which Gerard Hamilton was most apprehensive of his pupils not being sufficiently constant in the practice of, is misrepresentation. Under the name of *action*, thrice was gesticulation spoken of as the first accomplishment of his profession by the Athenian orator;

By Gerard Hamilton, in a collection of aphorisms, 553 in number, in about 40 vice is recommended without disguise; twelve times is misrepresentation, i. e. premeditated falsehood with or without a mask,

recommended in the several forms of which it presented itself to him as susceptible : viz. in the way of
false addition three times, in the way of false substitution twice, and in the way of omission seven times.

He was fearful of deceiving the only persons he
meant not to deceive (viz. the pupils to whom he was
teaching the art of deceiving others), had he fallen
into any such omission, as that of omitting in the
teaching of this lesson any instruction or example that
might contribute to render them perfect in it.

Of a good cause as such, of every cause that is entitled to the appellation of a good cause, it is the characteristic property that it does not stand in need,—of
a bad cause, of every cause that is justly designated
by the appellation of a bad cause, it is the characteristic property that it does stand in need of assistance
of this kind. Not merely indifference as between good
and bad, but predilection for what is bad is therefore
the cast of mind betrayed or rather displayed by Gerard Hamilton. For the praise of intelligence and
active talent, that is, for so much of it as constitutes
the difference between what is to be earned by the
advocation of good causes only, and that which is to
be earned by the advocation of bad causes likewise,
—of bad causes in preference to good ones,—for this
species and degree of praise it is, that Gerard Hamilton was content to forgo the merit of probity, of sincerity as a branch of probity, and take to himself the
substance as well as the shape and colour of the opposite vice.

This is the work which, having been fairly written out by the author [a], and thence by the editor, presumed to have been intended for the press, had been "shown by him to his friend Dr. Johnson." This is the work which this same Dr. Johnson, if the editor is to be believed, "considered a very curious and masterly performance." This is the work in which that pom-

---

[a] Extract from the preface to Hamilton's work:—

"But in the treatise on Parliamentary Logic we have the fruit and result of the experience of one, who was by no means unconversant with law, and had himself sat in Parliament for more than forty years; who in the commencement of his political career burst forth like a meteor, and for a while obscured his contemporaries by the splendour of his eloquence; who was a most curious observer of the characteristic merits and defects of the distinguished speakers of his time: and who, though after his first effort he seldom engaged in public debate, devoted almost all his leisure and thoughts, during the long period above mentioned, to the examination and discussion of all the principal questions agitated in Parliament, and of the several topics and modes of reasoning by which they were either supported or opposed.

"Hence the rules and precepts here accumulated, which are equally adapted to the use of the pleader and orator: nothing vague, or loose, or general *, is delivered; and the most minute particularities and artful turns of debate are noticed with admirable acuteness, subtilty and precision. The work, therefore, is filled with practical axioms, and parliamentary and forensic wisdom, and cannot but be of perpetual use to all those persons who may have occasion to use their discursive talents within or without the doors of the House of Commons, in conversation at the Bar, or in Parliament.

"This Tract was fairly written out by the author, and therefore may be presumed to have been intended by him for the press. He had shown it to his friend Dr. Johnson, who considered it a very curious and masterly performance."

---

* For "nothing," read "the greatest part." J. B.

pous preacher of melancholy moralities saw, if the editor is to be believed, nothing to " object to," but " the too great conciseness and refinement of some parts of it," and the occasion it gave to " a *wish* that some of the precepts had been more opened and expanded."

So far as concerns sincerity and candour in debate, the two friends indeed, even to judge of them from the evidence transmitted to us by their respective panegyrists, seem to have been worthy to smell at the same nosegay: and an " expansion and enlargement," composed by the hand that suggested it, would beyond doubt have been a " very curious and masterly," as well as amusing addition to this " very curious and masterly performance."

Two months before his death, when, if he himself is to be believed, ambition had in such a degree been extinguished in him by age and infirmities, that after near forty years of experience a seat in Parliament was become an object of indifference to him [a],—four years after he had been visited by a fit of the palsy [b],— he was visited by a fit of virtue, and in the paroxysm of that fit hazarded an experiment, the object of which was to try whether, in a then approaching parliament, a seat might not be obtained without a complete sacrifice of independence. The experiment was not successful. From some lord, whose name decorum has suppressed, he was, as his letter to his lordship

---

[a] page 26.   [b] page 14.

testified, "on the point of receiving" a seat; and the object of this letter was to learn whether, along with the seat, "the power of thinking for himself" might be included in the grant; the question being accompanied with a request that, in case of the negative, some other nominee might be the object of his lordship's "confidence."

The request was inadmissible, and the confidence found some other object.

It is in the hope of substituting men to puppets, and the will of the people to the will of noble lords, puppets themselves to ministers or secret advisers, that parliamentary reform has of late become once more an object of general desire: but parliamentary reform was that sort of thing which "he would sooner," he said, "suffer his hand to be cut off than vote for [a]:" whether it was before or after the experiment that this magnanimity was displayed, the editor has not informed us.

The present which the world received in the publication of this work may on several accounts be justly termed a valuable one. The only cause of regret is, that the editor should, by the unqualified approbation and admiration bestowed upon it, have made the principles of the work as it were his own.

True it is, that where instruction is given, showing how mischief may be done or aimed at, whether it shall serve as a precept or a prohibition, depends in

---

[a] page xxxvii.

the upshot, upon the person on whom it operates with effect :

Many a dehortation that not only has the effect of an exhortation, but was designed to have that effect ;

Instructions how to administer poisons with success, may on the other hand have the effect of enabling a person who takes them up with an opposite view, to secure himself the more effectually against the attack of poisons ;

But by the manner in which he writes, by the accessory ideas presented by the words in which the instruction is conveyed, there can seldom be much dificulty in comprehending in the delivery of his instructions whether the writer wishes that the suggestions conveyed by them should be embraced or rejected:

If occasionally there can be room for doubt in this respect, at any rate no room can there be for any in the case of Gerard Hamilton. As little can there be in the case of his editor and panegyrist: *Qui mihi discipulus puer es, cupis atque doceri, Huc ades, hæc animo concipe dicta tuo :* The object or end in view is, on occasion of a debate in Parliament,—in a supreme legislative assembly,—how to gain your point, whatever it be. The means indicated as conducive to that end are sometimes fair ones, sometimes foul ones ; and be they fair or foul, they are throughout delivered with the same tone of seriousness and composure.

Come unto me all ye who have a point to gain, and I will show you how : bad or good, so as it be not parliamentary reform, to me it is matter of indifference.

Here then, whatever be the influence of authority, authority in general, and that of the writer in particular, it is in the propagation of insincerity (of insincerity to be employed in the service it is most fit for, and in which it finds its richest reward,) that throughout the whole course of this work, and under the name of Gerard Hamilton, not to speak of his editor and panegyrist, such authority exerts itself.

To secure their children from falling into the vice of drunkenness, it was the policy we are told of Spartan fathers to exhibit their slaves in a state of inebriation, that the contempt might be felt to which a man stands exposed when the intellectual part of his frame has been thrown into the disordered state to which it is apt by this means to be reduced. An English father, if he has any regard for the morals of his son, and in particular for that vital part in which sincerity is concerned, will perhaps no where else find so instructive an example as Gerard Hamilton has rendered himself by this book: in that mirror may be seen to what a state of corruption the moral part of man's frame is capable of being reduced; to what a state of degradation, in the present state of parliamentary morality, a man is capable of sinking even when sober, and without any help from wine; and with what deliberate zeal he may himself exert his powers in the endeavour to propagate the infection in other minds.

PART THE FIRST.

FALLACIES OF AUTHORITY,

The subject of which is Authority in various shapes, and the object, to repress all exercise of the reasoning faculty.

WITH reference to any measures having for their object *the greatest happiness of the greatest number,* the course pursued by the adversaries of such measures has commonly been, in the first instance, to endeavour to repress altogether the exercise of the reasoning faculty, by adducing authority in various shapes as conclusive upon the subject of the measure proposed.

But before any clear view can be given of the deception liable to be produced by the abuse of the species of argument here in question, it will be necessary to bring to view the distinction between the proper and the improper use of it.

In the ensuing analysis of Authority, one distinction ought to be borne in mind;—it is the distinction between what may be termed a question of *opinion,* or *quid faciendum;* and what may be termed a question of *fact,* or *quid factum.* Since it will frequently happen, that whilst the authority of a person in respect to a question of fact is entitled to more or less regard, it is not so entitled in respect of a question of opinion.

CHAPTER I.

Sect. 1. *Analysis of Authority.*
Sect. 2. *Appeal to Authority, in what cases fallacious.*

I. WHAT on any given occasion is the legitimate weight or influence of authority regard being had to the different circumstances in which a person, the supposed declaration of whose opinion constitutes the authority in question, was placed at the time of the delivery of such declaration?

1st. Upon the degree of relative and adequate *intelligence* on the part of the person whose opinion or supposed opinion constitutes the authority in question, —say of the *persona cujus,* 2dly, Upon the degree of relative *probity* on the part of that same person, 3dly, Upon the nearness or remoteness of the relation between the immediate subject of such his opinion and the question in hand, 4thly, Upon the fidelity of the medium, through which such supposed opinion has been transmitted (including correctness and completeness),—upon such circumstances, the legitimately persuasive force of the authority thus constituted, seems to depend : such are the sources in which any deficiency in respect of such persuasive force is to be looked for.

Deficiency of attention, i. e. intensity and steadiness of attention with reference to the influencing circumstances on which the opinion in order to be correct, required to be grounded ; deficiency in respect of opportunity or matter of information, with reference to the individual question in hand ; distance in point of time from the scene of the proposed measure ; distance in point of place ;—such again are the sources in which, the situation of the person in question being given, any deficiency in respect of relative and adequate intelligence is, it seems, to be looked for.

It is in the character of a cause of deficiency in relative and adequate information, that distance in point of time operates as a cause of deficiency in respect of relative and adequate intelligence, and so in regard to distance in point of place.

As to relative probity, any deficiency referable to this head will be occasioned by the exposure of the *persona cujus* to the action of sinister interest : concerning which see Part 5, Chapter 2.—*Causes of the utterance of these fallacies.*

The most ordinary and conspicuous deficiency in the article of relative probity, is that of sincerity : the improbity consisting in the opposition or discrepancy between the opinion expressed and the opinion really entertained.

But as not only declaration of opinion, but opinion itself, is exposed to the action of sinister interest, in so far as this is the case, the deficiency is occasioned in two ways ; by the action of the sinister interest either the relevant means and materials are kept out of the mind, or, if this be not found practicable, the attention is kept from fixing upon them with the degree of intensity proportioned to their legitimately persuasive force.

As to the mass of information received by any person in relation to a given subject, the correctness and completeness of such information, and thence the probability of correctness on the part of the opinion grounded on it, will be in the joint ratio of the sufficiency of the *means* of collecting such information, and the strength of the *motives* by which he was urged to the employment of those means.

On both these accounts taken together, at the top of the scale of trustworthiness stands that mass of authority which is constituted by what may be termed *scientific* or professional *opinion:* that is, opinion entertained in relation to the subject in question by a person who, by special means and motives attached to a particular situation in life, may with reason be considered as possessed of such *means* of ensuring the correctness of his opinion as cannot reasonably be expected to have place on the part of a person not so circumstanced.

As to the special *motives* in question, they will in every case be found to consist of good or evil : profit for instance, or loss, presenting themselves as eventually likely to befall the person in question; profit or other good in case of the correctness of his opinion; loss or other evil in the event of its incorrectness.

In proportion to the force with which a man's will

is operated upon by the motives in question, is the degree of attention employed in looking out for the means of information, and the use made of them in the way of reflection towards the formation of his opinion.

Thus in the case of every occupation which a man engages in with a view to profit, the hope of gaining his livelihood, and the fear of not gaining it, are the motives by which he is urged to apply his attention to the collection of whatsoever information may contribute to the correctness of the several opinions which he may have occasion to form, respecting the most advantageous method of carrying on the several operations, by which such profit may be obtained.

1. The legitimately persuasive force of professional authority, being taken as the highest term in the scale, the following may be noticed as expressive of so many other species of authority, occupying so many inferior degrees in the same scale :

2. Authority derived from *power*. The greater the quantity of power a man has, no matter in what shape, the nearer the authority of his opinion comes to professional authority, in respect of the facility of obtaining the means conducive to correctness of decision.

3. Authority derived from *opulence*. Opulence —being an instrument of power, and, to a considerable extent, applicable in a direct way to many or most of the purposes to which power is applicable,—

seems to stand next after power in the scale of instruments of facility as above.

4. Authority derived from *reputation*, considered as among the efficient causes of respect. By reputation, understand, on this occasion, general reputation, not special and relative reputation, which would rank the species of authority under the head of professional authority as above.

Note, that of all these four species of authority it is only in the case of the first that the presumable advantage, which is the efficient cause of its legitimately persuasive force, extends to the article of motives as well as means. By having the *motives* that tend to correctness of information, the professional man has the *means* likewise; since it is to the force of the motives under the stimulus of which he acts that he is indebted for whatever means he acquires. It is from his having the motives that it follows that he has the means.

But in those other cases, whatsoever be the *means* which a man's situation places within his reach, it follows not that he has the *motives*,—that he is actually under the impulse of any motive sufficient to the full action of that desire and that energy by which alone he can be in an adequate degree put in possession of the means.

On the contrary, in proportion as in the scale of power the man in question rises above the ordinary level, in that same proportion, in respect of motives for exertion (be the line of action what it may), he is apt to sink below the same level : because the greater the quantum of the share of the general mass of objects of desire that a man is already in possession of, the greater is the amount of that portion of his desires which is already in a state of saturation, and consequently the less the amount of that portion which, remaining unsatiated, is left free to operate upon his mind in the character of a motive.

Under Oriental despotism, the person at whose command the *means* of information exist in a larger proportion than they do in the instance of any other person whatever, is the *despot ;* but necessary *motives* being wanting, no use is made by him of these means, and the general result is a state of almost infantine imbecility and ignorance.

Such in kind, varying only in degree, is the case with every hand in which power is lodged, unincumbered with obligation ; or, in other words, with sense of eventual danger.

In England the king, the peer, the opulent borough-holding or county-holding country gentleman, should, on the above principle, present an instance of the sort of double scale in question, in which, while means decrease, motives rise.

But so long as he takes any part at all in public affairs, the sense of that weak kind of eventual responsibility to which, notwithstanding the prevailing habits of idolatry, the monarch, as such, stands at all times exposed, suffices to keep his intellectual faculties at a

84553

point more or less above the point of utter ignorance;
whereas, short of provable idiotism, there is no degree
of imbecility that in either of those two other situations can suffice to render it matter of danger or inconvenience to the possessor, either to leave altogether unexercised the power annexed to such situation, or, without the smallest regard for the public
welfare, to exercise it in whatever manner may be
most agreeable or convenient to himself.

All this while, it is only on the supposition of perfect relative probity, viz. of that branch of probity that
consists of sincerity, as well as absence of all such
sources of delusion as to the person in question are
liable to produce the effects of insincerity,—in a word,
it is only on the supposition of the absence of exposure
to the action of any sinister interest, operating in such
direction as to tend to produce either erroneous opinion or misrepresentation of a man's opinion on the subject in question, that, in so far as it depends on the information necessary to correctness of opinion, the title
of a man's authority to regard bears any proportion
either to motives or to means of information as above.

On the contrary, if either immediately, or through
the medium of the will, a man's understanding be exposed to the dominion of sinister interest, the more
complete as well as correct the mass of relative information is which he possesses, the more completely
destitute of all title to regard, i. e. to confidence, unless it be in the opposite direction, will the authority,
or pretended or real opinion, be.

Hence it is that on the question, What is the system of remuneration best adapted to the purpose of obtaining the highest degree of official aptitude throughout the whole field of official service ?—the authority of any person who here or elsewhere, now or formerly, was in possession or expectation of any such situation as that of minister of state, so far from being greater than that of an average man, is not equal to 0, but in the mathematical sense negative, or so much below 0 : i. e. so far as it affords a reason for looking upon the opposite opinion as the right and true one.

So again as to this question—What, in so far as concerns cognoscibility, or economy and expedition in procedure, the state of the law *ought* to be ?—in the instance of any person who here or elsewhere, recently or formerly, but more particularly in this country, was in possession or expectation of any situation, professional or official, the profitableness of which, in the shape of pecuniary emolument, or in any other shape (such as power, reputation, ease, and occasionally vengeance), depended upon the incognoscibility, the expensiveness, the dilatoriness, the vexatiousness of the system of judicial procedure,—the weight of the authority,—the strength of its title to credit on the part of those understandings to which the force of it is applied,—is not merely equal to 0, but in the mathematical sense negative, or so much below 0.

Note, that where, as above, the weight or probative force of the authority in question is spoken of as being not positive but negative (being rendered so by sinister interest), what is taken for granted is, that the direction in which the authority is offered is the same as that in which the sinister interest acts; for if, the direction in which the sinister interest acts lying one way, the direction in which the opinion acts lies the other way, in such case the title of the opinion to credit on the part of the understandings to which it is proposed, so far from being destroyed or weakened, is much increased, because the grounds for correctness of opinion, the motives and the means which in that case lead to correctness being more completely within the reach of, and according to probability present to, the minds of this class of men, the forces that tend to promote aberration having by this supposition spent themselves in vain, the chance for correctness is thereby greater.

Accordant with this, and surely enough accordant with experience and common sense, is one of the few rational rules that as yet have received admittance among the technically established rules of evidence. In a man's own favour his own testimony is the weakest, —in his disfavour, the strongest, evidence.

It is on this account that, wherever a man is in a superior degree furnished as above with means of, and motives for, obtaining relevant information, the stronger the force of the sinister interest under the action of which his opinion is delivered, the stronger is his title to attention. In the way of direct and relevant argument applying to the question in hand in a direct and specific way, if the question be suscep-

tible of any such arguments, in proportion to the efficiency of the motives and means he has for the acquisition of such relevant information is the probability
of his bringing such information to view. If, then,
instead of bringing to view any such relevant information, or by way of supplement and support to such
relevant information (when weak and insufficient), the
arguments which he brings to view are of the irrelevant
sort, the addition of such bad arguments affords a sort
of circumstantial evidence, and that of no mean degree of probative force, of the inability of the side thus
advocated to furnish any good ones.

Closeness of the relation between the immediate
subject in hand and the subject of the supposed opinion of which the authority is composed, has been
mentioned as the third circumstance necessary to be
considered in estimating the credit due to authority :
—of this, it is evident enough, there cannot be any
common and generally applicable measure. It is that
sort of quantity of the amount of which a judgment
can only be pronounced in each individual case.

As to the fourth,—fidelity of the *medium* through
which the opinion constitutive of the authority in question has been, or is supposed to have been, transmitted,
—it is only *pro memoria* that this topic is here brought
to view in the list of the circumstances from which
the legitimately persuasive force of an opinion constitutive of authority is liable to experience decrease :
of its admission into this list the propriety is, on the

bare mention, as manifest as it is in the power of reasoning to make it.   In this respect the rule and measure as well as cause of such decrease stand exactly
on the same ground as the rule with respect to any
other evidence ; authority being to the purpose in
question neither more nor less than an article of circumstantial evidence.

The need for the legitimately persuasive force of
authority, i. e. probability of comparatively superior
information on the one hand, is in the inverse ratio of
information on the part of the person on whom it is
designed to operate, on the other.   The less the degree in which each man is qualified to form a judgment on any subject on the ground of specific and relevant information,—on the ground of direct evidence,
—the more cogent the necessity he is under of trusting,
with a degree of confidence more or less implicit, to
that species of circumstantial evidence : and in proportion to the number of the persons who possess,
each within himself, the means of forming an opinion on any given subject on the ground of such
direct evidence, the greater the number of the persons to whom it ought to be matter of shame to
frame and pronounce their respective decision, on no
better ground than that of such inconclusive and necessarily fallacious evidence.

Of the truth of this observation, men belonging to
the several classes, whose situation in the community
has given to them in conjunction, with efficient power,

a separate and sinister interest opposite to that of the community in general, have seldom failed to be in a sufficient degree percipient.

In this perception, in the instance of the fraternity of lawyers, may be seen one cause, though not the only one, of the anxiety betrayed, and pains taken, to keep the rule of action in a state of as complete incognoscibility as possible on the part of those whose conduct is professed to be directed by it, and whose fate is in fact disposed of by it.

In this same perception, in the instance of the clergy of old times in the Romish church, may be seen in like manner the cause, or at least one cause, of the pains taken to keep in the same state of incognoscibility the acknowledged rule of action in matters of sacred and supernatural law.

In this same perception, in the instance of the English clergy of times posterior to those of the Romish church,—in this same perception,—may be seen one cause of the exertions made by so large a proportion of the governing classes of that hierarchy to keep back and if possible render abortive the system of invention, which has for its object the giving to the exercise of the art of reading the highest degree of universality possible.

To return. Be the subject matter what it may, to the account of fallacies cannot be placed any mention made of an opinion to such or such an effect, as having been delivered or intimated by such or such a person by name, when the sole object of the reference is to point out a place where relevant arguments adduced on a given occasion may be found in a more complete or perspicuous state than they are on the occasion on which they are adduced.

In the case thus supposed there is no *irrelevancy.* The arguments referred to are, by the supposition, relevant ones; such as, if the person by whom they have been presented to view were altogether unknown, would not lose any thing of their weight: the opinion is not presented as constitutive of authority, as carrying any weight of itself, and independently of the considerations which he has brought to view.

Neither is there any fallacy in making reference to the opinion of this or that professional person, in a case to such a degree professional or scientific, with relation to the hearers or readers, that the forming a correct judgment on such relevant and specific arguments as belong to it, is beyond their competence. In matters touching medical science, chemistry, astronomy, the mechanical arts, the various branches of the art of war, &c., no other course could be pursued.

Sect. 2. *Appeal to authority, in what cases fallacious* [a].

The case in which reference to authority is open to the imputation of fallacy, is where, in the course of a debate touching a subject lying in such sort within

---

[a] " An unquestionable maxim " (it is said) is this :—" Reason and not authority should determine the judgement:" said? and by whom? even by a bishop; and by what bishop? even Bishop Warburton : and

the comprehension of the debaters, that argument
bearing the closest relation to it would be perfectly
within the sphere of their comprehension,—*authority*
(a sort of argument in the case here in question not
relevant) is employed in the place of such relevant
arguments as might have been adduced on one side,
or, in opposition to irrelevant ones, adduced on the
other side.

But the case in which the practice of adducing authority in the character of an argument is in the highest
degree exposed to the imputation of fallacy, is, where
the situation of the debaters being such that the forming a correct conception of, and judgment on, such
relevant arguments as the subject admits is not beyond
their competency, the *opinion*, real or supposed, of any
person who, from his profession or other particular
situation, derives an interest opposite to that of the
public, is adduced in the character of an argument, in
lieu of such relevant arguments as the question ought
to furnish. (In an appendix to this chapter will be
given examples of persons whose declared opinions,
on a question of legislation, are in a peculiar degree
liable to be tinged with falsity by the action of sinister
interest.)

He who, on a question concerning the propriety of
any law or established practice with reference to the
time being, refers to authority as decisive of the ques-

---

this not in one work only, but in two. The above words are from his
Div. Legat. 2, 302; and in his Alliance, &c. is a passage to the same
effect : here then we have authority against authority.

tion, assumes the truth of one or other of two positions : viz. that the principle of utility, i. e. that the greatest happiness of the greatest number, is not at the time in question the proper standard for judging of the merits of the question, or that the practice of other and former times, or the opinion of other persons, ought to be regarded in all cases as conclusive evidence of the nature and tendency of the practice :—conclusive evidence, superseding the necessity and propriety of any recourse to reason or present experience.

In the first case, being really an enemy to the community, that he should be esteemed as such by all to whom the happiness of the community is an object of regard, is no more than right and reasonable, no more than what, if men acted consistently, would uniformly take place.

In the other case, what he does, is, virtually to acknowledge himself not to possess any powers of reasoning which he himself can venture to think it safe to trust to : incapable of forming for himself any judgment by which he looks upon it as safe to be determined, he betakes himself for safety to some other man, or set of men, of whom he knows little or nothing, except that they lived so many years ago; that the period of their existence was by so much anterior to his own time ; by so much anterior, and consequently possessing for its guidance so much the less experience.

But when a man gives this account of himself,—

when he represents his own mind as labouring under this kind and degree of imbecility,—what can be more reasonable than that he should be taken at his word? that he should be considered as a person labouring under a general and incurable imbecility, from whom nothing relevant can reasonably be expected?

He who, in place of reasoning, deduced (if the subject be of a practical nature) from the consideration of the end in view, employs authority, makes no secret of the opinion he entertains of his hearers or his readers : he assumes that those to whom he addresses himself are incapable, each of them, of forming a judgment of their own. If they submit to this insult, may it not be presumed that they acknowledge the justice of it?

Of imbecility, at any rate of self-conscious and self-avowed imbecility, proportionable humility ought naturally to be the result ;

On the contrary, so far from humility,—of this species of idolatry,—of this worshipping of dead men's bones, all the passions the most opposite to humility, —pride, anger, obstinacy, and overbearingness,—are the frequent, not to say the constant accompaniments.

With the utmost strength of mind that can be displayed in the field of reasoning, no reasonable man ever manifests so much heat, assumes so much, or exhibits himself disposed to bear so little, as these men, whose title to regard and notice is thus given up by themselves.

Whence this inconsistency? Whence this violence?

From this alone, that having some abuse to defend, some abuse in which they have an interest and a profit, and finding it on the ground of present public interest indefensible, they fly for refuge to the only sort of argument, in which so much as the pretension of being sincere in error can find countenance.

By authority, support, the strength of which is proportioned to the number of the persons joining in it, is given to systems of opinions, at once absurd and pernicious—to the religion of Buddh, of Brama, of Foh, of Mahomet.

And hence it may be inferred that the probative force of authority is not increased by the number of those who may have professed a given opinion, unless indeed it could be proved that each individual of the multitudes who professed the opinion, possessed in the highest degree the means and motives for ensuring its correctness. Even in such a case it would not warrant the substitution of the authority for such direct evidence and arguments as any case in debate might be able to supply, supposing the debaters capable of comprehending such direct evidence and arguments; but that, in ordinary cases, no such circumstantial evidence should possess any such legitimately probative force as to warrant the addition, much less the substitution of it, to that sort of information which belongs to direct evidence, will, it is supposed, be rendered sufficiently apparent by the following considerations ;

1. If in theory any the minutest degree of force

were ascribed to the elementary monade of the body of authority thus composed, and this theory were followed up in practice, the consequence would be, the utter subversion of the existing state of things :—as for example,— If distance in point of time were not sufficient to destroy the probative force of such authority, the Catholic religion would in England be to be restored to the exclusive dominion it possessed and exercised for so many centuries : the Toleration laws would be to be repealed, and persecution to the length of extirpation would be to be substituted to whatever liberty in conduct and discourse is enjoyed at present; —and in this way, after the abolished religion had thus been triumphantly restored, an inexorable door would be shut against every imaginable change in it, and thence against every imaginable reform or improvement in it, through all future ages :

2. If distance in point of place were not understood to have the same effect, some other religion than the Christian,—the religion of Mahomet for example, or the way of thinking in matters of religion, prevalent in China,—would have to be substituted by law to the Christian religion.

In authority, defence, such as it is, has been found for every imperfection, for every abuse, for every the most pernicious and most execrable abomination that the most corrupt system of government has ever husbanded in its bosom :—

And here may be seen the mischief necessarily attached to the course of him whose footsteps are regulated by the finger of this blind guide.

What is more, from hence may inferences be deduced—nor those ill-grounded ones—respecting the probity or improbity, the sincerity or insincerity, of him who, standing in a public situation, blushes not to look to this blind guide, to the exclusion of, or in preference to, reason—the only guide that does not begin with shutting his own eyes, for the purpose of closing the eyes of his followers.

As the world grows older, if at the same time it grows wiser, (which it will do unless the period shall have arrived at which experience, the mother of wisdom, shall have become barren,) the influence of authority will in each situation, and particularly in Parliament, become less and less.

Take any part of the field of moral science, private morality, constitutional law, private law,—go back a few centuries, and you will find argument consisting of reference to authority, not exclusively, but in as large a proportion as possible. As experience has increased, authority has been gradually set aside, and reasoning, drawn from facts and guided by reference to the end in view, true or false, has taken its place.

Of the enormous mass of Roman law heaped up in the school of Justinian,—a mass, the perusal of which would employ several lives occupied by nothing else, —materials of this description constitute by far the greater part. A. throws out at random some loose

thought : B., catching it up, tells you what A. thinks
—at least, what A. said : C. tells you what has been
said by A. and B.; and thus like an avalanche the
mass rolls on.

Happily it is only in matters of law and religion
that endeavours are made, by the favour shown and
currency given to this fallacy, to limit and debilitate
the exercise of the right of private inquiry in as great
a degree as possible, though at this time of day the
exercise of this essential right can no longer be suppressed in a complete and direct way by legal punishment.

In mechanics, in astronomy, in mathematics, in
the new-born science of chemistry,—no one has at this
time of day either effrontery or folly enough to avow,
or so much as to insinuate, that the most desirable
state of these branches of useful knowledge, the most
rational and eligible course, is to substitute decision
on the ground of authority, to decision on the ground
of direct and specific evidence.

In every branch of physical art and science, the
folly of this substitution or preference is matter of
demonstration,—is matter of intuition, and as such is
universally acknowledged.   In the moral branch of
science, religion not excluded, the folly of the like
receipt for correctness of opinion would not be less
universally recognised, if the wealth, the ease, and the
dignity attached to and supported by the maintenance
of the opposite opinion, did not so steadily resist such
recognition.

Causes of the employment and prevalence of this fallacy.

It is obvious that this fallacy, in all its branches, is so frequently resorted to by those who are interested in the support of abuses, or of institutions pernicious to the great body of the people, with the intention of suppressing all exercise of reason. A foolish or untenable proposition resting on its own support, or the mere credit of the utterer, could not fail speedily to encounter detection and exposure;—the same proposition extracted from a page of Blackstone, or from the page or mouth of any other person to whom the idle and unthinking are in the habit of unconditionally surrendering their understandings, shall disarm all opposition.

Blind obsequiousness, ignorance, idleness, irresponsibility, anticonstitutional dependence, anticonstitutional independence, are the causes which enable this fallacy to maintain such an ascendancy in the governing assemblies of the British empire.

First, In this situation one man is on each occasion ready to borrow an opinion of another, because through ignorance and imbecility he feels himself unable, or through want of solicitude unwilling, to form one for himself; and he is thus ignorant, if natural talent does not fail him, because he is so *idle.* Knowledge, especially in so wide and extensive a field, requires study; —study, labour of mind bestowed with more or less energy, for a greater or less length of time.

But, Secondly, In a situation for which the strongest talents would not be more than adequate, there is frequently a failure of natural talent; because, in so many instances admission to that situation depends either on the person admitted, or on others to whom, whether he has or has not the requisite talents is a matter of indifference, that no degree of intellectual deficiency, short of palpable idiocy, can have the effect of excluding a man from occupying it.

Thirdly, The sense of responsibility is in the instance of a large proportion of the members wanting altogether; because in so small a proportion are they at any time in any degree of dependence on the people whose fate is in their hands, and because in the instance of the few who are in any degree so dependent, the efficient cause and consequently the feeling of such dependence endures during so small a proportion of the time for which they enjoy their situations : because also, while so few are dependent on those on whom they ought to be dependent, so many are dependent on those who ought to be dependent on them, —those servants of the crown, on whose conduct they are commissioned by their constituents to act as judges. What share of knowledge, intelligence and natural talent is in the house, is thus divided between those who are, and their rivals who hope to be, servants of the crown. The consequence is, that, those excepted in whom knowledge, intelligence and talent are worse than useless, the house is composed of men the furniture of whose minds is made up of discordant prejudices, of which on each occasion they follow that by which the interest or passion of the moment is most promoted.

Then, with regard to responsibility, so happily have matters been managed by the house,—a seat there is not less clear of obligation than a seat in the opera house : in both, a man takes his seat, then only when he cannot find more amusement elsewhere ; for both, the qualifications are the same,—a ticket begged or bought : in neither is a man charged with any obligation, other than the negative one of not being a nuisance to the company; in both, the length as well as number of attendances depends on the amusement a man finds, except, in the case of the house, as regards the members dependent on the crown. True it is, that a self-called independent member is not necessarily ignorant and weak : if by accident a man possessed of knowledge and intelligence is placed in the house, his seat will not deprive him of his acquirements : all therefore that is meant is, only, that ignorance does not disqualify, not that knowledge does. Of the crown and its creatures it is the interest that this ignorance be as thick as possible. Why ? Because the thicker the ignorance, the more completely is the furniture of men's minds made up of those interest-begotten prejudices, which render them blindly obsequious to all those who with power in their hands stand up to take the lead.

But the emperor of Morocco is not more irresponsible, and therefore more likely to be ignorant and prone to be deceived by the fallacy of authority, than a member of the British Parliament :—the emperor of Morocco's power is clear of obligation ; so is the member's :—the emperor's power, it is true, is an integer, and the member's but a fraction of it ; but no ignorance prevents a man from becoming or continuing emperor of Morocco, nor from becoming or continuing a member :—the emperor's title is derived from birth ; so is that of many a member :—to enjoy his despotism, no fraud, insincerity, hypocrisy or jargon is necessary to the emperor ; much of all to the member :—by ascending and maintaining his throne, no principle is violated by the emperor ; by the member, if a borough-holder, many are violated on his taking and retaining his seat :—by being a despot, the emperor is not an impostor ; the member is :—the emperor pretends not to be a trustee, agent, deputy, delegate, representative ; lying is not among the accompaniments of his tyranny and insolence ; the member does pretend all this, and (if a borough-holder) lies.—A trust-holder? yes; but a trust-breaker : —an agent? yes ; but for himself :—a representative of the people? yes; but so as Mr. Kemble is of Macbeth :—a deputy? yes ; because it has not been in their power to depute, to delegate any body else :— deputy,—delegate,—neither title he assumes but for argument, and when he cannot help it ; deputation being matter of fact, the word presents an act with all its circumstances, viz. fewness of the electors, their want of freedom, &c. ; representation is a more con-

venient word, the acts, &c. are kept out of sight by it ;
—it is a mere fiction, the offspring of lawyer-craft, and
any one person or thing may be represented by any
other. By canvass with colours, a man is represented ; by a king, the whole people ; by an ambassador,
the king, and thus the people.

Remedy against the influence of this fallacy.

For banishing ignorance, for substituting to it a
constantly competent measure of useful, appropriate
and general instruction, the proper, the necessary, the
only means lie not deep beneath the surface.

The sources of instruction being supposed at command, and the quantity of natural talent given, the
quantity of information obtained will in every case be
as the quantity of mental labour employed in the collection of it—the quantity of mental labour, as the
aggregate strength of the motives by which a man is
excited to labour.

In the existing order of things, there is, comparatively speaking, no instruction obtained, because no
labour is bestowed,—no labour is bestowed, because
none of the motives by which men are excited to labour are applied in this direction.

The situation being by the supposition an object of
desire, if the case were such that, without labour employed in obtaining instruction, there would be no
chance of obtaining the situation, or but an inferior
chance, while in case of labour so employed there
would be a certainty or a superior chance,—here, in-

struction would have its motives,—here, labour applied to the attainment of instruction,—here, consequently, instruction itself would have its probably efficient cause.

The quality, i. e. the relative applicability of the mass of information obtained, is an object not to be overlooked.

The goodness of the quality will depend on the liberty enjoyed in respect of the choice. By prohibitions, with penalties attached to the delivery of alleged information relative to a subject in question, or any part of it, the quality of the whole mass is impaired, and an implied certificate is given of the truth and utility of whatsoever portion is thus endeavoured to be suppressed.

APPENDIX.

Examples of descriptions of persons whose declared opinions upon a question of legislation are peculiarly liable to be tinged with falsity by the action of sinister interest.

1. Lawyers; oppositeness of their interest to the universal interest.

The opinions of lawyers in a question of legislation, particularly of such lawyers as are or have been practising advocates, are peculiarly liable to be tinged with falsity by the operation of sinister interest. To the interest of the community at large, that of every advocate is in a state of such direct and constant opposition (especially in civil matters), that the above assertion requires an apology to redeem it from the appearance of trifling : the apology consists in the extensively prevailing propensity to overlook and turn aside from a fact so entitled to notice. It is the people's interest that delay, vexation and expense of procedure should be as small as possible :—it is the advocate's that they should be as great as possible : viz. expense in so far as his profit is proportioned to it ; factitious vexation and delay, in so far as inseparable from the profit-yielding part of the expense. As to uncertainty in the law, it is the people's interest that each man's security against wrong should be as complete as possible ; that all his rights should be known to him ; that all acts, which in the case of his doing them will be treated as offences, may be known to him as such, together with their eventual punishment, that he may avoid committing them, and that others may, in as few instances as possible, suffer either from the wrong or from the expensive and vexatious remedy. Hence it is their interest, that as to all these matters the rule of action, in so far as it applies to each man, should at all times be not only discoverable, but actually present to his mind. Such knowledge, which it is every man's interest to possess to the greatest, it is the lawyer's interest that he possess it to the narrowest extent possible. It is every man's interest to keep out of lawyers' hands as much as possible ; it is the lawyer's interest to get him in as often, and keep him in as long, as possible : thence that any written expression of the words necessary to keep non-lawyers out of his hand may as long as possible be prevented from coming into

existence, and when in existence as long as possible
kept from being present to his mind, and when presented from staying there [a]. It is the lawyer's interest,
therefore, that people should continually suffer for the
non-observance of laws, which, so far from having received efficient promulgation, have never yet found
any authoritative expression in words. This is the
perfection of oppression : yet, propose that access to
knowledge of the laws be afforded by means of a code,
lawyers, one and all, will join in declaring it impossible. To any effect, as occasion occurs, a judge will
forge a rule of law : to that same effect, in any determinate form of words, propose to make a law, that
same judge will declare it impossible. It is the judge's
interest that on every occasion his declared opinion
be taken for the standard of right and wrong; that
whatever he declares right or wrong be universally received as such, how contrary soever such declaration
be to truth and utility, or to his own declaration at
other times :—hence, that within the whole field of
law, men's opinions of right and wrong should be as
contradictory, unsettled, and thence as obsequious to
him as possible : in particular, that the same conduct

---

[a] A considerable proportion of what is termed the Common law of
England is in this oral and unwritten state. The cases in which it
has been clothed with words, that is, in which it has been framed and
pronounced, are to be found in the various collections of reported decisions. These decisions, not having the sanction of a law passed by
the legislature, are confirmed or overruled at pleasure by the existing
judges ; so that, except in matters of the most common and daily occurrence, they afford no rule of action at all.

which to others would occasion shame and punishment, should to him and his occasion honour and reward : that on condition of telling a lie, it should be
in his power to do what he pleases, the injustice and
falsehood being regarded with complacency and re-
verence ; that as often as by falsehood, money or advantage in any other shape can be produced to him,
it should be regarded as proper for him to employ reward or punishment, or both, for the procurement of
such falsehood.   Consistently with men's abstaining
from violences, by which the person and property of
him and his would be alarmingly endangered, it is his
interest that intellectual as well as moral depravation
should be as intense and extensive as possible ; That
transgressions cognizable by him should be as numerous as possible ; That injuries and other transgressions committed by him should be reverenced as
acts of virtue ; That the suffering produced by such
injuries should be placed, not to his account, but
to the immutable nature of things, or to the wrong-
doer, who, but for encouragement from him, would not
have become such. His professional and personal interest being thus adverse to that of the public, from
a lawyer's declaration that the tendency of a proposed
law relative to procedure, &c. is pernicious, the contrary inference may not unreasonably be drawn. From
those habits of misrepresenting their own opinion
(i. e. of insincerity), which are almost peculiar to this
in comparison with other classes, one presumption is,
that he does not entertain the opinion thus declared ;

—another, that if he does, he has been deceived into it by sinister interest and the authority of co-professional men, in like manner deceivers or deceived : in other words, it is the result of interest-begotten prejudice.   In the case of every other body of men, it is generally expected that their conduct and language will be for the most part directed by their own interest, that is, by their own view of it.   In the case of the lawyer, the ground of this persuasion, so far from being weaker, is stronger than in any other case.   His evidence being thus *interested evidence,* according to his own rules his declaration of opinion on the subject here pointed out would not be so much as hearable.   It is true, were those rules consistently observed, judicature would be useless, and society dissolved : accordingly they are not so observed, but observed or broken pretty much at pleasure ; but they are not the less among the number of those rules, the excellence and inviolability of which the lawyer is never tired of trumpeting.   But on any point, such as those in question, nothing could be more unreasonable, nothing more inconsistent with what has been said above, than to refuse him a *hearing.*   On every such point, his habits and experience afford him facilities not possessed by any one else for finding relevant and specific arguments, when the nature of the case affords any ; but the surer he is of being able to find such arguments, if any such are to be found, the stronger the reason for treating his naked declaration of opinion as unworthy of all regard : accompanied by specific arguments, it is useless ; desti-

tute of them, it amounts to a virtual confession of
their non-existence.

So matters stand on the question what *ought* to be law.

On the question what the law *is*, so long as the rule of
action is kept in the state of common, alias unwritten,
alias imaginary law, authority, though next to nothing,
is every thing. The question is, what on a given occasion A. (the judge) is likely to think: wait till your
fortune has been spent in the inquiry, and you will
know ; but, forasmuch as it is naturally a man's wish
to be able to give a guess what the result will eventually be, before he has spent his fortune, in the
view if possible to avoid spending his fortune and get-
ting nothing in return for it, he applies through the
medium of B. (an attorney) for an opinion to C. (a
counsel), who, considering what D. (a former judge)
has, on a subject supposed to be more or less analogous to the one in question, said or been supposed to
say, deduces therefore his guess as to what, when the
time comes, Judge A., he thinks, will say, and gives it
you. A shorter way would be to put the question at
once to A. ; but, for obvious reasons, this is not permitted.

On many cases, again, as well-grounded a guess
might be had of an astrologer for five shillings, as of a
counsel for twice or thrice as many guineas, but that
the lawyer considers the astrologer as a smuggler, and
puts him down.

But Packwood's opinion on the goodness of his own
razors would be a safer guide for judging of their good-

ness, than a judge's opinion on the goodness of a proposed law : it is Packwood's interest that his razors
be as good as possible ;—the judge's, that the law *be*
as bad, yet *thought to be* as good, as possible.   It
would not be the judge's interest that his commodity
should be thus bad, if, as in the case of Packwood, the
customer had other shops to go to ; but in this case,
even when there are two shops to go to, the shops
being in confederacy, the commodity is equally bad in
both ; and the worse the commodity, the better it is
said to be.   In the case of the judge's commodity, no
experience suffices to undeceive men ; the bad quality
of it is referred to any cause but the true one.

Example 2. Churchmen; oppositeness of their interest
to the universal interest.

In the lawyer's case it has been shown that on the
question, what on such or such a point *ought* to be
law, to refer to a lawyer's opinion given without or
against specific reasons, is a fallacy ; its tendency, in
proportion to the regard paid to it, deceptious ;—the
cause of this deceptious tendency, sinister interest, to
the action of which all advocates and (being made
from advocates) all judges stand exposed.   To the
churchman's case the same reasoning applies : as, in
the lawyer's case the objection does not arise on the
question, what law *is*, but what *ought to be* law,—so
in the churchman's case it does not arise as to what
in matters of religion is law, but as to what in those
matters ought to be law.   On a question not connected

with religion, reference to a churchman's opinion as *such*, as authority, can scarcely be considered as a fallacy, such opinion not being likely to be considered as constitutive of authority. To understand how great would be the probability of deception, if on the question, what in matters of religion ought to be law, the unsupported opinion of a churchman were to be regarded as authority, we must develop the nature and form of the sinister interest, by which any declaration of opinion from such a quarter is divested of all title to regard. The sources of a churchman's sinister interest are as follows :—

1. On entering into the profession, as condition precedent to advantage from it in the shape of subsistence and all other shapes, he makes of necessity a solemn and recorded declaration of his belief in the truth of 39 articles, framed 262 years ago, the date of which, the ignorance and violence of the time considered, should suffice to satisfy a reflecting mind of the impossibility of their being all of them really believed by any person at present :

2. In this declaration is generally understood to be included an engagement or undertaking, in case of original belief and subsequent change, never to declare, but, if questioned, to deny such change :

3. In the institution thus established, he beholds shame and punishment attached to sincerity, rewards in the largest quantity to absurdity and insincerity. Now the presumptions resulting from such an application of reward and punishment to engage men to declare as-

sent to given propositions are, 1st, That the proposition is not believed by the proposer; 2nd, Thence, that it is not true; 3rd, Thence, that it is not believed by the acceptor. It is impossible by reward or punishment to produce real and immediate belief: but the following effects may certainly be produced: 1st, The abstaining from any declaration of disbelief; 2nd, Declaration of belief; 3rd, The turning aside from all considerations tending to produce disbelief; 4th, The looking out for, and fastening exclusive attention to, all considerations tending to produce belief, authority especially, by which a sort of vague and indistinct belief of the most absurd propositions has every where been produced.

On no other part of the field of knowledge are reward or punishment now-a-days considered as fit instruments for the production of assent or dissent. A schoolmaster would not be looked upon as sane, who, instead of putting Euclid's Demonstrations into the hands of his scholar, should, without the Demonstrations, put the Propositions into his hand, and give him a guinea for signing a paper declarative of his belief in them, or lock him up for a couple of days without food on his refusal to sign it. And so in chemistry, mechanics, husbandry, astronomy, or any other branch of knowledge. It is true, that in those parts of knowledge in which assent and dissent are left free, the importance of truth may be esteemed not so great as here, where it is thus influenced; but the more important the truth, the more flagrant the absurdity and

F

tyranny of employing, for the propagation of it, instruments, the employment of which has a stronger
tendency to propagate error than truth.

4. For teaching such religious truths as men are
allowed to teach, together with such religious error as
they are thus forced to teach, the churchman sees rewards allotted in larger quantities than are allotted to
the most useful services. Of much of the matter of
reward thus bestowed, the disposal is in the king's
hands, with the power of applying it, and motives for
applying it, to the purpose of parliamentary service,
paying for habitual breach of trust, and keeping in
corrupt and secret dependence on his agents, those
agents of the people whose duty it is to sit as judges
over the agents of the king. In Ireland, of nine-tenths
of those on pretence of instructing whom this vast mass
of reward is extorted, it is known, that, being by conscience precluded from hearing, it is impossible that
they should derive any benefit from such instruction.

In Scotland, where Government reward is not employed in giving support to it, Church-of-Englandism
is reduced to next to nothing.

The opinions which, in this state of things, interest
engages a churchman to support, are—1st, That reward to the highest extent has no tendency to promote insincerity, even where practicable, to an unlimited extent, and without chance of detection; 2nd,
Or that money given in case of compliance, refused
in case of non-compliance, is not reward for compliance; 3rd, Or that punishment applied in case of

non-compliance, withheld in case of compliance, is not punishment; 4th, Or that insincerity is not vice but virtue, and as such ought to be promoted ; 5th, That it is not merely consistent with, but requisite to, good government to extort money from poor and rich to be applied as reward for doing nothing, or for doing but a small part of that which is done by others for a small proportion of the same reward, and this on pretence of rendering service, which nine-tenths of the people refuse to receive.

It is the interest of the persons thus engaged in a course of insincerity, that by the same means perseverance in the same course should be universal and perpetual ; for suppose, in case of the reward being withheld, the number annually making the same declaration should be reduced to half : this would be presumptive evidence of insincerity on the part of half of those who made it before.

The more flagrant the absurdity, the stronger is each man's interest in engaging as many as possible in joining with him in the profession of assent to it ; for the greater the number of such co-declarants, the greater the number of those of whose professions the elements of authority are composed ; and of those who stand precluded from casting on the rest the imputation of insincerity.

The following, then, are the abuses in the defence of which all churchmen are enlisted : 1st. Perpetuation of immorality in the shape of insincerity ; 2. Of absurdity in subjects of the highest importance ;

3. Extortion inflicted on the many for the benefit of the few ; 4. Reward bestowed on idleness and incapacity to the exclusion of labour and ability ; 5. The matter of corruption applied to the purposes of corruption in a constant stream ; 6. In one of these kingdoms a vast majority of the people kept in degradation avowedly for no other than the above purposes. But whoever is engaged by interest in the support of any one Government abuse, is engaged in the support of all, each giving to the others his support in exchange.

It being the characteristic of abuse to need and receive support from fallacy, it is the interest of every man who derives profit from abuse in any shape to give the utmost currency to fallacy in every shape, viz. as well those which render more particular service to others' abuses as those which render such service to his own.   It being the interest of each person so situated to give the utmost support to abuse, and the utmost currency to fallacy in every shape, it is also his interest to give the utmost efficiency to the system of education by which men are most effectually divested both of the power and will to detect and expose fallacies, and thence to suppress every system of education in proportion as it has a contrary tendency : lastly, the stronger the interest by which a man is urged to give currency to fallacy, and thus to propagate deception, the more likely is it that such will be his endeavour : the less fit, therefore, will his opinion be to serve in the character of authority, as a standard and model for the opinions of others.

CHAPTER II.

The wisdom of our ancestors; or Chinese argument.

Ad verecundiam.

Sect. 1. *Exposition.*

THIS argument consists in stating a supposed repugnancy between the proposed measure and the opinions of men by whom the country of those who are
discussing the measure was inhabited in former times;
these opinions being collected either from the express
words of some writer living at the period of time in
question, or from laws or institutions that were then
in existence.

*Our wise ancestors—the wisdom of our ancestors—
the wisdom of ages—venerable antiquity—wisdom of
old times—*

· Such are the leading terms and phrases of propositions the object of which is to cause the alleged
repugnance to be regarded as a sufficient reason for
the rejection of the proposed measure.

Sect. 2. *Exposure.*

This fallacy affords one of the most striking of the
numerous instances in which, under the conciliatory
influence of *custom*, that is of *prejudice*, opinions
the most repugnant to one another are capable of
maintaining their ground in the same intellect.

This fallacy, prevalent as it is in matters of law, is directly repugnant to a principle or maxim universally admitted in almost every other department of human intelligence, and which is the foundation of all useful knowledge and of all rational conduct.

" Experience is the mother of wisdom," is among the maxims handed down to the present and all future ages, by the wisdom, such as it has been, of past ages.

No! says this fallacy, the true mother of wisdom is, not *experience*, but *inexperience*.

An absurdity so glaring carries in itself its own refutation ; and all that we can do is, to trace the causes which have contributed to give to this fallacy such an ascendancy in matters of legislation.

Among the several branches of the fallacies of authority, the cause of delusion is more impressive in this than in any other.

1st, From inaccuracy of conception arises incorrectness of expression ; from which expression, conception, being produced again, error, from having been a momentary cause, comes to be a permanent effect.

In the very denomination commonly employed to signify the portion of time to which the fallacy refers, is virtually involved a false and deceptious proposition, which, from its being employed by every mouth, is at length, without examination, received as true.

What in common language is called *old* time, ought (with reference to any period at which the fallacy in question is employed) to be called *young* or early time.

As between individual and individual living at the same time and in the same situation, he who is old, possesses, as such, more experience than he who is young ;—as between generation and generation, the reverse of this is true, if, as in ordinary language, a preceding generation be, with reference to a succeeding generation, called *old ;*—the *old* or preceding generation could not have had so much experience as the succeeding. With respect to such of the materials or sources of wisdom which have come under the cognisance of their own senses, the two are on a par ;— with respect to such of those materials and sources of wisdom as are derived from the reports of others, the later of the two possesses an indisputable advantage.

In giving the name of old or elder to the earlier generation of the two, the misrepresentation is not less gross, nor the folly of it less incontestable, than if the name of old man or old woman were given to the infant in its cradle.

What then is the wisdom of the times called old ? Is it the wisdom of gray hairs? No.—It is the wisdom of the cradle [a].

The learned and honourable gentlemen of THIBET

---

[a] No one will deny that preceding ages have produced men eminently distinguished by benevolence and genius; it is to them that we owe in succession all the advances which have hitherto been made in the career of human improvement : but as their talents could only be developed in proportion to the state of knowledge at the period in which they lived, and could only have been called into action with a view to then-existing circumstances, it is absurd to rely on their authority, at a period and under a state of things altogether different.

do homage to superior wisdom—superiority raised to
the degree of divinity—in the person of an infant
lying and squalling in his cradle.

The learned and honourable gentlemen of WEST-
MINSTER set down as impostors the LAMAS of THI-
BET, and laugh at the folly of the deluded people
on whom such imposture passes for sincerity and
wisdom.

But the worship paid at THIBET to the infant body
of the present day, is, if not the exact counterpart, the
type at least of the homage paid at WESTMINSTER
to the infant minds of those who have lived in earlier
ages.

2ndly, Another cause of delusion which promotes
the employment of this fallacy, is the reigning prejudice in favour of the dead ;—a prejudice which,
in former times, contributed, more than any thing
else, to the practice of idolatry : the dead were speedily
elevated to the rank of divinities ; the superstitious
invoked them, and ascribed a miraculous efficacy to
their relics.

This prejudice, when examined, will be seen to be
no less indefensible than pernicious—no less pernicious than indefensible.

By propagating this mischievous notion, and acting
accordingly, the man of selfishness and malice obtains
the praise of humanity and social virtue. With this
jargon in his mouth, he is permitted to sacrifice the
real interests of the living to the imaginary interests
of the dead. Thus imposture, in this shape, finds

the folly or improbity of mankind a never-failing fund
of encouragement and reward.

De mortuis nil nisi bonum ;—with all its absurdity,
the adage is but too frequently received as a leading
principle of morals.   Of two attacks, which is the
more barbarous, on a man that does feel it, or on a
man that does not ?   On the man that does feel it, says
the principle of utility : On the man that does not,
says the principle of caprice and prejudice—the principle of sentimentalism—the principle in which imagination is the sole mover—the principle in and by
which feelings are disregarded as not worth notice.

The same man who bepraises you when dead, would
have plagued you without mercy when living.

Thus as between Pitt and Fox.   While both were
living, the friends of each reckoned so many adversaries in the friends of the other.   On the death of him
who died first, his adversaries were converted into
friends.   At what price this friendship was paid for
by the people is no secret [a] : see the Statute Book, see
the debates of the time, and see *Defence of Economy*
against Burke and Rose.

The cause of this so extensively-prevalent and ex-
tensively-pernicious propensity lies not very deep.

A dead man has no rivals,—to nobody is he an object of envy,—in whosesoever way he may have stood
when living, when dead he no longer stands in any

---

[a] For the payment of Mr. Pitt's creditors was voted 40,000*l.* of the
public money :—to Mr. Fox's widow, 1500*l.* a year.

body's way. If he was a man of genius, those who denied him any merit during his life, even his very enemies, changing their tone all at once, assume an air of justice and kindness, which costs them nothing, and enables them, under pretence of respect for the dead, to gratify their malignity towards the living.

Another class of persons habitually exalts the past for the express purpose of depressing and discouraging the present generation.

It is characteristic of the same sort of persons, as well as of the same system of politics, to idolize, under the name of wisdom of our ancestors, the wisdom of untaught inexperienced generations, and to undervalue and cover with every expression of contempt that the language of pride can furnish, the supposed ignorance and folly of the great body of the people [a].

So long as they keep to vague generalities,—so long as the two objects of comparison are each of them taken in the lump,—wise ancestors in one lump, ignorant and foolish mob of modern times in the other, —the weakness of the fallacy may escape detection. Let them but assign for the period of superior wisdom any determinate period whatsoever, not only will the groundlessness of the notion be apparent (class being compared with class in that period and the present one), but, unless the antecedent period be comparatively speaking a very modern one, so wide will be the disparity, and to such an amount in favour of modern

------

[a] A " Burdett mob," for example.

times, that, in comparison of the lowest class of the
people in modern times (always supposing them pro-
ficients in the art of reading, and their proficiency employed in the reading of newspapers), the very highest
and best informed class of these wise ancestors will
turn out to be grossly ignorant.

Take for example any year in the reign of Henry
the Eighth, from 1509 to 1546. At that time the
House of Lords would probably have been in possession of by far the larger proportion of what little
instruction the age afforded : in the House of Lords,
among the laity, it might even then be a question
whether without exception their lordships were all of
them able so much as to read. But even supposing
them all in the fullest possession of that useful art,
political science being the science in question, what
instruction on the subject could they meet with at
that time of day ?

On no one branch of legislation was any book extant from which, with regard to the circumstances of
the then present times, any useful instruction could
be derived : distributive law, penal law, international
law, political economy, so far from existing as sciences,
had scarcely obtained a name : in all those departments,
under the head of *quid faciendum,* a mere blank : the
whole literature of the age consisted of a meagre
chronicle or two, containing short memorandums of
the usual occurrences of war and peace, battles,
sieges, executions, revels, deaths, births, processions,
ceremonies, and other external events; but with scarce

a speech or an incident that could enter into the composition of any such work as a history of the human mind,—with scarce an attempt at investigation into causes, characters, or the state of the people at large. Even when at last, little by little, a scrap or two of political instruction came to be obtainable, the proportion of error and mischievous doctrine mixed up with it was so great, that whether a blank unfilled might not have been less prejudicial than a blank thus filled, may reasonably be matter of doubt.

If we come down to the reign of James the First, we shall find that Solomon of his time, eminently eloquent as well as learned, not only among crowned but among uncrowned heads, marking out for prohibition and punishment the practices of devils and witches, and without any the slightest objection on the part of the great characters of that day in their high situations, consigning men to death and torment for the misfortune of not being so well acquainted as he was with the composition of the Godhead.

Passing on to the days of Charles the Second, even after Bacon had laid the foundations of a sound philosophy, we shall find Lord Chief Justice Hale (to the present hour chief god of the man of law's idolatry) unable to tell (so he says himself) what *theft* was; but knowing at the same time too well what witchcraft was, hanging men with the most perfect complacency for both crimes, amidst the applauses of all who were wise and learned in that blessed age.

Under the name of Exorcism the Catholic liturgy

contains a form of procedure for driving out devils :—
even with the help of this instrument, the operation
cannot be performed with the desired success but by
an operator qualified by holy orders for the working
of this as well as so many other wonders.

In our days and in our country the same object is
attained, and beyond comparison more effectually, by
so cheap an instrument as a common newspaper : before this talisman, not only devils but ghosts, vam-
pires, witches, and all their kindred tribes, are driven
out of the land, never to return again ; the touch of
holy water is not so intolerable to them as the bare
smell of printers' ink.

If it is absurd to rely on the wisdom of our ancestors, it is not less so to vaunt their probity : they were
as much inferior to us in that point as in all others ;
and the further we look back, the more abuses we
shall discover in every department of Government :—
nothing but the enormity of those abuses has produced
that degree of comparative amendment on which at
present we value ourselves so highly. Till the human
race was rescued from that absolute slavery under
which nine-tenths of every nation groaned, not a single step could be made in the career of improvement;
and take what period we will in the lapse of preceding
ages, there is not one which presents such a state of
things as any rational man would wish to see entirely re-established.

Undoubtedly, the history of past ages is not wanting in some splendid instances of probity and self-de-

votion ; but in the admiration which these excite, we commonly overrate their amount, and become the dupes of an illusion occasioned by the very nature of an extensive retrospect.   Such a retrospect is often made by a single glance of the mind ; in this glance the splendid actions of several ages (as if for the very purpose of conveying a false estimate of their number and contiguity) present themselves, as it were, in a lump, leaving the intervals between them altogether unnoticed.   Thus groves of trees, which at a distance present the appearance of thick and impenetrable masses, turn out on nearer approach to consist of trunks widely separated from each other.

Would you then have us speak and act as if we had never had any ancestors ?   Would you, because recorded experience, and, along with it, wisdom, increases from year to year, annually change the whole body of our laws?   By no means : such a mode of reasoning and acting would be more absurd even than that which has just been exposed ; and *provisional* adherence to existing establishments is grounded on considerations much more rational than a reliance on the wisdom of our ancestors.   Though the *opinions* of our ancestors are as such of little value, their *practice* is not the less worth attending to ; that is, in so far as their practice forms part of our own experience. However, it is not so much from what they did, as from what they underwent (good included as well as evil), that our instruction comes.   Independently of consequences, what they did is no more than evidence of what they thought; nor yet, in legislation, is it evidence of what they thought best for the whole community, but only of what the rulers thought would be best for themselves in periods when every species of abuse prevailed unmitigated, by the existence of either public press or public opinion. From the facts of their times, much information may be derived :—from the opinions, little or none. As to opinions, it is rather from those which were foolish than from those which were well grounded, that any instruction can be derived. From foolish opinions comes foolish conduct; from the most foolish conduct, the severest disaster; and from the severest disaster, the most useful warning. It is from the folly, not from the wisdom, of our ancestors that we have so much to learn; and yet it is to their wisdom, and not to their folly, that the fallacy under consideration sends us for instruction.

It seems, then, that our ancestors, considering the disadvantages under which they laboured, could not have been capable of exercising so sound a judgment on their interests as we on ours: but as a knowledge of the facts on which a judgment is to be pronounced is an indispensable preliminary to the arriving at just conclusions, and as the relevant facts of the later period must all of them individually, and most of them specifically, have been unknown to the man of the earlier period, it is clear that any judgment derived from the authority of our ancestors, and applied to existing affairs, must be a judgment pronounced without evidence; and this is the judgment which the fallacy in question calls on us to abide by, to the exclusion

of a judgment formed on the completest evidence that
the nature of each case may admit.

Causes of the Propensity to be influenced by this Fallacy.

Wisdom of ancestors being the most impressive of
all arguments that can be employed in defence of
established abuses and imperfections, persons interested in this or that particular abuse are most forward to
employ it.

But their exertions would be of little avail, were it
not for the propensity which they find on the part of
their antagonists to attribute to this argument nearly
the same weight as those by whom it is relied on.

This propensity may be traced to two intimately-
connected causes:—1. Both parties having been trained up alike in the school of the English lawyers, headed
by Blackstone; and, 2. Their consequent inability, for
want of practice, to draw from the principle of general utility the justificative reason of every thing that is
susceptible of justification.

In the hands of a defender of abuse, authority answers a double purpose, by affording an argument in
favour of any particular abuse which may happen to
call for protection, and by causing men to regard with
a mingled emotion of hatred and terror the principle
of general utility, in which alone the true standard
and measure of right and wrong is to be found.

In no other department of the field of knowledge
and wisdom (unless that which regards religion be an
exception) do leading men of the present times recom-

mend to us this receipt for thinking and acting wisely. By no gentleman, honourable or right honourable, are we sent at this time of day to the wisdom of our ancestors for the best mode of marshalling armies, navigating ships, attacking or defending towns; for the best modes of cultivating and improving land, and preparing and preserving its products for the purposes of food, clothing, artificial light and heat; for the promptest and most commodious means of conveyance of ourselves and goods from one portion of the earth's surface to another; for the best modes of curing, alleviating or preventing disorders in our own bodies and those of the animals which we contrive to apply to our use.

Why this difference? Only because in any other part of the field of knowledge, legislation excepted, (and religion, in so far as it has been taken for the subject of legislation,) leading men are not affected with that sinister interest which is so unhappily combined with power in the persons of those leading men who conduct Governments as they are generally at present established.

Sir *H. Davy* has never had any thing to gain, either from the unnecessary length, the miscarriage, or the unnecessary part of the expenses attendant on chemical experiments; he therefore sends us either to his own experiments or to those of the most enlightened and fortunate of his cotemporaries, and not to the notions of Stahl, Van Helmont, or Paracelsus.

G

CHAPTER III.

1. *Fallacy of Irrevocable Laws.*
2. *Fallacy of Vows.*

Ad superstitionem.

THE two fallacies brought to view in this chapter
are intimately connected, and require to be considered
together : the object in view is the same in both, the
difference lies only in the instrument employed ; and
both of them are in effect the fallacy of *the wisdom of
our ancestors,* pushed to the highest degree of extra-
vagance and absurdity.

The object is to tie up the hands of future legislators
by obligations supposed to be indissoluble.

In the case of the fallacy derived from the alleged
irrevocable nature of certain laws, or, to speak briefly,
the fallacy of *Irrevocable laws*, the instrument employed is a contract—a contract entered into by the
ruling powers of the state in question with the ruling
powers of some other party. This other party may be
either the sovereign of some other state, or the whole
or some part of the people of the state in question.

In the case of the fallacy derived from *vows*, a supernatural power is called in and employed in the
character of guarantee.

Fallacy of Irrevocable Laws.

Exposition.

A law, no matter to what effect, is proposed to a
legislative assembly, and, no matter in what way, it is

by the whole or a majority of the assembly regarded
as being of a beneficial tendency.   The fallacy in
question consists in calling upon the assembly to reject it notwithstanding, upon the single ground, that
by those who, in some former period, exercised the
power which the present assembly is thus called on to
exercise, a regulation was made, having for its object
the precluding for ever, or to the end of a period not
yet expired, all succeeding legislators from enacting a
law to any such effect as that now proposed.

What will be tolerably clear to every man who will
allow himself to think it so, is—that, notwithstanding
the profound respect we are most of us so ready to
testify towards our fellow creatures as soon as the
moment has arrived after which it can be of no use to
them, the comforts of those who are out of the way of
all the comforts we can bestow, as well as of all the
sufferings we can inflict, are not the real objects to
which there has been this readiness to sacrifice the
comforts of present and future generations, and that
therefore there must be some other interest at the
bottom.

Exposure.

1. To consider the matter in the first place on the
ground of general utility.

At each point of time the sovereign for the time
possesses such means as the nature of the case affords
for making himself acquainted with the exigencies of
his own time.

With relation to the future, the sovereign has no such means of information ; it is only by a sort of vague anticipation, a sort of rough and almost random guess drawn by analogy, that the sovereign of this year can pretend to say what will be the exigencies of the country this time ten years.

Here then, to the extent of the pretended immutable law, is the government transferred from those who possess the best possible means of information, to those who, by their very position, are necessarily incapacitated from knowing any thing at all about the matter.

Instead of being guided by their own judgment, the men of the 19th century shut their own eyes, and give themselves up to be led blindfold by the men of the 18th century.

The men who have the means of knowing the whole body of the facts on which the correctness and expediency of the judgment to be formed, must turn, give up their own judgment to that of a set of men entirely destitute of any of the requisite knowledge of such facts.

Men who have a century more of experience to ground their judgments on, surrender their intellect to men who had a century less experience, and who, unless that deficiency constitutes a claim, have no claim to preference.

If the prior generation were, in respect of intellectual qualification, ever so much superior to the subsequent generation,—if it understood so much better than the subsequent generation itself the interest of that subsequent generation,—could it have been in an equal degree anxious to promote that interest, and consequently equally attentive to those facts with which, though in order to form a judgment it ought to have been, it is impossible that it should have been acquainted? In a word, will its love for that subsequent generation be quite so great as that same generation's love for itself?

Not even here, after a moment's deliberate reflection, will the assertion be in the affirmative.

And yet it is their prodigious anxiety for the welfare of their posterity that produces the propensity of these sages to tie up the hands of this same posterity for evermore, to act as guardians to its perpetual and incurable weakness, and take its conduct for ever out of its own hands.

If it be right that the conduct of the 19th century should be determined not by its own judgment but by that of the 18th, it will be equally right that the conduct of the 20th century should be determined not by its own judgment but by that of the 19th.

The same principle still pursued, what at length would be the consequence?—that in process of time the practice of legislation would be at an end: the conduct and fate of all men would be determined by those who neither knew nor cared any thing about the matter; and the aggregate body of the living would remain for ever in subjection to an inexorable tyranny, exercised, as it were, by the aggregate body of the dead.

This irrevocable law, whether good or bad at the moment of its enactment, is found at some succeeding period to be productive of mischief—uncompensated mischief—to any amount. Now of this mischief, what possibility has the country of being rid?

A despotism, though it were that of a Caligula or a Nero, might be to any degree less mischievous, less intolerable, than any such immutable law. By benevolence (for even a tyrant may have his moments of benevolence), by benevolence, by prudence,—in a word, by caprice,—the living tyrant might be induced to revoke his law, and release the country from its consequences. But the dead tyrant! who shall make *him* feel? who shall make *him* hear?

Let it not be forgotten, that it is only to a bad purpose that this and every other instrument of deception will in general be employed.

It is only when the law in question is mischievous, and generally felt and understood to be such, that an argument of this stamp will be employed in the support of it.

Suppose the law a good one, it will be supported, not by absurdity and deception, but by reasons drawn from its own excellence.

But is it possible that the restraint of an irrevocable law should be imposed on so many millions of living beings by a few score, or a few hundreds, whose existence has ceased? Can a system of tyranny be established under which the living are all slaves—and a few among the dead, their tyrants?

The production of any such effect in the way of constraint being physically impossible, if produced in any degree it must be by force of argument—by the force of fallacy, and not by that of legislative power.

The means employed to give effect to this device may be comprised under two heads; the first of them exhibiting a contrivance not less flagitious than the position itself is absurd.

1st, In speaking of a law which is considered as repugnant to any law of the pretended immutable class, the way has been to call it void. But to what purpose call it void? Only to excite the people to rebellion in the event of the legislator's passing any such void law. In speaking of a law as void, either this is meant or nothing. It is a sophism of the same cast as that expressed by the words *rights of man*, though played off in another shape, by a different set of hands, and for the benefit of a different class.

Are the people to consider the law void? They are then to consider it as an act of injustice and tyranny under the name of law;—as an act of power exercised by men who have no right to exercise it : they are to deal by it as they would by the command of a robber; they are to deal by those who, having passed it, take upon them to enforce the execution of it, as they would deal, whenever they found themselves strong enough, by the robber himself [a].

---

[a] Lord Coke was for holding void every act contrary to Magna Charta. If his doctrine were tenable, every act imposing law-taxes would be void.

2ndly, The other contrivance for maintaining the immutability of a given law, is derived from the notion of a contract or engagement. The faithful observance of contracts being one of the most important of the ties that bind society together, an argument drawn from this source cannot fail to have the appearance of plausibility.

But be the parties interested who they may, a contract is not itself an end; it is but a means toward some end : and in cases where the public is one of the parties concerned, it is only in so far as that end consists of the happiness of the whole community, taken in the aggregate, that such contract is worthy to be observed.

Let us examine the various kinds of contract to which statesmen have endeavoured to impart this character of perpetuity :—1, Treaties between state and foreign state, by which each respectively engages its government and people; 2, Grant of privileges from the sovereign to the whole community in the character of subjects; 3, Grant of privileges from the sovereign to a particular class of subjects ; 4, New arrangement of power between different portions or branches of the sovereignty, or new declaration of the rights of the community ; 5, Incorporative union between two sovereignties having or not having a common head.

Take, then, for the subject and substance of the contract any one of these arrangements : so long as the happiness of the whole community, taken in the aggregate, is in a greater degree promoted by the exact observance of the contract than it would be by any alteration, exact ought to be the observance :—on the contrary, if, by any given change, the aggregate of happiness would be in a greater degree promoted than by the exact observance, such change ought to be made.

True it is, that, considering the alarm and danger which is the natural result of every breach of a contract to which the sovereignty is party, in case of any change with respect to such contract, the aggregate of public happiness will be in general rather diminished than promoted, unless, in case of disadvantage produced to any party by the change, such disadvantage be made up by adequate compensation.

Let it not be said that this doctrine is a dangerous doctrine, because the compensation supposed to be stipulated for as adequate may prove but a nominal, or at best but an inadequate, compensation. Reality and not pretence, probity not improbity, veracity not mendacity, are supposed alike on all sides ;—the contract a real contract, the change a real change, the compensation an adequate as well as real compensation. Instead of probity suppose improbity in the sovereignty; it will be as easy to deny the existence, or explain away the meaning of the contract, or to deny or explain away the change, as, instead of a real to give a nominal, instead of an adequate to give an inadequate, compensation.

To apply the foregoing principles to the cases above enumerated, one by one.

1. In the case of the contract or treaty between state and foreign state, the dogma of immutability has seldom been productive of any considerable practical inconvenience: the ground of complaint has arisen rather from a tendency to change than a too rigid adherence to the treaty.

However, some commercial treaties between state and state, entered into in times of political ignorance or error, and pernicious to the general interests of commerce, are frequently upheld under a pretence of regard for the supposed inviolability of such contracts, but in reality from a continuance of the same ignorance, error, antipathy or sinister interest, which first occasioned their existence. It can seldom or never happen that a forced direction thus given to the employment of capital can ultimately prove advantageous to either of the contracting parties; and when the pernicious operation of such a treaty on the interests of both parties has been clearly pointed out, there can be no longer any pretence for continuing its existence. Notice, however, of any proposed departure from the treaty ought to be given to all the parties concerned; sufficient time should be afforded to individuals engaged in traffic, under the faith of the treaty, to withdraw, if they please, their capitals from such traffic, and in case of loss, compensation as far as possible ought to be afforded.

2. Grant of privilege from the sovereign to the whole community in the character of subjects.—If, by the supposed change, privileges to equal value be given

in the room of such as are abrogated, adequate compensation is made : if greater privileges are substituted, there is the greater reason for supporting the measure.

3. Grant of privileges from the sovereign to a particular class of subjects.

No such particular privilege ought to have been granted if the aggregate happiness of the community was likely to be thereby diminished : but, unless in case of a revocation, adequate compensation be here also made, the aggregate happiness of the community will not be increased by the change; the happiness of the portion of the community to be affected by the change, being as great a part of the aggregate happiness as that of any other portion of equal extent.

Under this head are included all those more particular cases in which the sovereign contracts with this or that individual, or assemblage of individuals, for money or money's worth, to be supplied, or service otherwise to be rendered.

4. New arrangement or distribution of powers as between different portions or branches of the sovereignty, or new declaration of the rights of the community.

Let the supposition be, that the result will not be productive of a real addition to the aggregate stock of happiness on the part of the whole community, it ought not to be made : let the supposition be the reverse, then, notwithstanding the existence of the contract, the change is such as it is right and fitting should be made.

The first of these can never furnish a case for compensation, unless in so far as, without charge or disadvantage to the people, the members of the sovereignty can contrive to satisfy one another; such members of the sovereignty being, as to the rest of the community, not proprietors but trustees.

The frame or constitution of the several American united states, so far from being declared immutable or imprescriptible, contains an express provision, that a convention shall be holden at intervals for the avowed object of revising and improving the constitution, as the exigencies of succeeding times may require. In Europe, the effect of declaring this or that article in a new distribution of powers, or in the original frame of a constitution, immutable, has been to weaken the sanction of all laws. The article in question turns out to be mischievous or impracticable; instead of being repealed, it is openly or covertly violated; and this violation affords a precedent or pretext for the non-observance of arrangements clearly calculated to promote the aggregate happiness of the community.

5. Case of an incorporative union between two sovereignties, having or not having a common head.

Of all the cases upon the list, this is the only one which is attended with difficulty.

This is the case in which, at the same time that a contract with detailed clauses is at once likely and fit to be insisted on, compensation, that compensation without which any change would not be consistent with general utility in the shape of justice or in any

other shape, is an operation attended with more difficulty than in any other of these cases.

Distressing indeed would be the difficulty, were it
not for one circumstance which happily is interwoven
in the very nature of the case.

At the time of the intended union, the two states
(not to embarrass the case by taking more than two
at a time) are, with relation each to the other, in a
greater or less degree foreign and independent states.

Of the two uniting states, one will generally be
more, the other less, powerful. If the inequality be
considerable, the more powerful state, naturally speaking, will not consent to the union, unless, after the
union, the share it possesses in the government of the
new-framed compound state be greater by a difference
bearing some proportion to the difference in prosperity
between the two states.

On the part of the less powerful state, precautions
against oppression come of course.

Wherever a multitude of human beings are brought
together, there is but too much room for jealousy, suspicion, and mutual ill-will.

In the apprehension of each, the others, if they obtain possession of the powers exercised by the common government, will be supposed to apply them unjustly. In men or in money, in labour or in goods,
in a direct way or in some indirect one, it may be the
study of the new compound government, under the
influence of that part of the quondam government
which is predominant in it, to render the pressure of

the contributions proportionably more severe upon the one portion of the new compounded state than upon the other, or to force upon it new customs, new religious ceremonies, new laws.

Let the hands of the new government remain altogether loose, one of the two compound nations may be injured and oppressed by the other.

Tie up the hands of the government in such degree as is requisite to give to each nation a security against injustice at the hands of the other, sooner or later comes the time in which the inconveniencies resulting from the restriction will become intolerable to one or other, or to both.

But sooner or later the very duration of the union produces the natural remedy.

Sooner or later, having for such or such a length of time been in the habit of acting in subjection to one government, the two nations will have become melted into one, and mutual apprehensions will have been dissipated by conjunct experience.

All this while, in one or both of the united states, the individuals will be but too numerous and too powerful who, by sinister interest and interest-begotten prejudice, will stand engaged to give every possible countenance and intensity to those fears and jealousies, to oppose to the entire composure of them every degree of retardation.

If, in either of the united communities at the time of the union, there existed a set of men more or less numerous and powerful, to whom abuse or imperfection in any shape was a source of profit, whatsoever restrictions may have been expressed in the contract, these restrictions will of course be laid hold of by the men thus circumstanced, and applied as far as possible to the giving protection and continuance to a state of things agreeable or beneficial to themselves.

At the time of the union between England and Scotland, the Tory party, of whom a large proportion were Jacobites, and all or most of them high-churchmen, had acquired an ascendant in the House of Commons.

Here, then, a favourable occasion presented itself to these partisans of Episcopacy for giving perpetuity to the triumph they had obtained over the English presbyterians, by the Act of Uniformity proclaimed in the time of Charles the Second[a].

In treaties between unconnected nations, where an advantage in substance is given to one, for the purpose of saving the honour of the other, it has been the custom to make the articles bear the appearance of reciprocity upon the face of them; as if, the facilitating the vent of French wines in England being the object of a treaty, provision were made in it that wine of the growth of either country might be imported into the other, duty free.

By the combined *astutia* of priestcraft and lawyer-craft, advantage was taken of this custom to rivet for ever those chains of ecclesiastical tyranny which, in

---

[a] 13 and 14 Charles II. c. 4.

the precipitation that attended the Restoration, had been fastened upon the people of England.—For securing the 45 Scotch members from being outnumbered by the 513 English ones, provision had been made in favour of the church of Scotland : therefore, on the principle of reciprocity for securing the 513 English members from being outnumbered by the 45 Scotch ones, like provision was made in favour of the church of England.

Blackstone avails himself of this transaction for giving perpetuity to whatever imperfections may be found in the ecclesiastical branch of the law, and the official establishment of England.

On a general account which he has been giving[a] of the articles and act of union, he grounds three observations :—

1. That the two kingdoms are now so inseparably united that nothing can ever disunite them again, except the mutual consent of both, or the successful resistance of either, upon apprehending an infringement of those points which, when they were separate and independent nations, it was mutually stipulated should be "fundamental and essential conditions of the union."

2. That, whatever else may be deemed "fundamental and essential conditions," the preservation of the two churches of England and Scotland, in the same state that they were in at the time of the union,

---

[a] Vol. i. 97, 98.

and the maintenance of the acts of uniformity which establish our common prayer, are expressly declared so to be.

3. That therefore any alteration in the constitution of either of those churches, or in the Liturgy of the church of England (unless with the consent of the respective churches collectively or representatively given), would be an infringement of these "fundamental and essential conditions," and greatly endanger the union.

On the original device, an improvement has, we see, been made by the ingenuity of the orthodox and learned commentator. If,—as for example by the alteration of any of the 39 articles,—if, by the abolition of any of the English ecclesiastical sinecures, or by any efficient measure for ensuring the performance of duty in return for salary, the ecclesiastical branch of the English official establishment were brought so much the nearer to what it is in Scotland, the Scotch, fired by the injury done to them, would cry out, a breach of faith! and call for a dissolution of the union.

To obviate this *danger*, a *great* one he denominates it, his ingenuity, in concert with his piety, has however furnished us with an expedient :—"The consent of the church collectively or representatively given," is to be taken; by which is meant, if any thing, that by the revival of the convocation, or some other means, the clergy of England are to be erected into a fourth estate.

H

What is evident is, that, unless the sinister influence of the Crown could be supposed to become *felo de se*, and employ itself in destroying a large portion of itself, nothing but a sincere persuasion of the utility of a change in relation to any of the points in question, and that entertained by a large proportion of the English members in each house, could ever be productive of any such change ;—that, in any attempt to force the discipline of the church of Scotland upon the church of England, the 45 Scotch members in the House of Commons, supposing them all unanimous, would have to outnumber, or some how or other to subdue, the 513 English ones ;—that in the House of Lords, the sixteen Scotch members, supposing all the lay lords indifferent to the fate of the church of England, would in like manner have to outnumber the 26 bishops and archbishops.

But the Tories, who were then in vigour, feared that they might not always be so, and seized that opportunity to fetter posterity by an act which should be deemed irrevocable.

The "administration of justice in Scotland [a]."—This forms the subject of the 19th article, which has for its *avowed* object the securing the people of Scotland against any such encroachments as might otherwise be made by the lawyers of England, by the use of those fictions and other frauds, in the use of which they had been found so expert. But throughout the

---

[a] 5 Ann. c. 8. art. 19. anno 1708.

whole course of this long article, the most rational and uniform care is taken to avoid all such danger as that of depriving the people of Scotland of such benefit as, from time to time, they might stand a chance of receiving at the hands of the united Parliament, by improvements in the mode of administering justice : "subject to such regulations as shall be made by the Parliament of Great Britain," is a clause over and over again repeated.

It would have been better for Scotland if, on the subject of the next article, viz. " heritable offices," including "heritable jurisdictions," the like wisdom had presided.   By that short article, those public trusts, together with others therein mentioned, are on the footing of "rights of property" reserved to the owners; yet still, without any expression of that fanatic spirit which, on the field of religion, had in the same statute occupied itself in the endeavour to invest the conceits of mortal man with the attribute of immortality.

Nine-and-thirty years after, came the act [a] for abolishing these same heritable jurisdictions.   Here was an act made in the very teeth of the act of union.

Mark now the sort of discernment, or of sincerity, that is to be learnt from Blackstone.

In a point blank violation of the articles of union, in the abolition of those heritable jurisdictions which

---

[a] "abolishing the heritable jurisdictions in Scotland" are so many words that stand in the title of it. Anno 1747, 20 Geo. 2. c. 43.

it was the declared object of one of its articles (20) to preserve, he saw nothing to "*endanger the union.*"

But suppose any such opinion to prevail, as that it is not exactly true that by the mere act of being born every human being merits *damnation* [a] (if by damnation be meant everlasting torment, or punishment in any other shape), and a corresponding alteration were made in the set of propositions called the 39 articles, the union would be "greatly endangered."

Between 20 and 30 years afterwards, at the suggestion of an honest member of the Court of Session, came upon the carpet, for the first time, the idea of applying remedies to some of the most flagrant imperfections in the administration of Scottish justice : and thereupon came out a pamphlet from James Boswell, declaiming, in the style of school-boy declamation, on the injury that would be done to the people of Scotland by rendering justice, or what goes by that name, a little less inaccessible to them, and the breach that would be made in the faith plighted by that treaty, which, to judge from what he says of it, he had never looked at.

Again, in 1806, when another demonstration was made of applying a remedy to the abuses and imperfections of the system of judicature in Scotland, every thing that could be done in that way was immediately reprobated by the Scotch lawyers as an infringement

---

[a] Art. 9.

of that most sacred of all sacred bonds—the union : nor, for the support of the brotherhood on the other side of the Tweed, was a second sight of the matter in the same point of view wanting in England.

As to any such design as that of oppressing their fellow subjects in Scotland, nothing could be further from the thoughts of the English members ; neither for good nor for evil uses was any expense of thought bestowed upon the matter. The ultimate object, as it soon became manifest, was the adding an item or two to the list of places.

Upon the whole, the following is the conclusion that seems to be dictated by the foregoing considerations. Every arrangement by which the hands of the sovereignty for the time being are attempted to be tied up, and precluded from giving existence to a fresh arrangement, is absurd and mischievous ; and, on the supposition that the utility of such fresh arrangement is sufficiently established, the existence of a prohibitive clause to the effect in question ought not to be considered as opposing any bar to the establishment of it.

True it is, that all laws, all political institutions, are essentially dispositions for the future ; and the professed object of them is, to afford a steady and permanent security to the interests of mankind. In this sense, all of them may be said to be framed with a view to perpetuity ; but *perpetual* is not synonymous with *irrevocable;* and the principle on which all laws ought to be, and the greater part of them have been, established, is that of *defeasible perpetuity;* a perpetuity defeasible only by an alteration of the circumstances and reasons on which the law is founded.

To comprise all in one word—Reason, and that alone, is the proper anchor for a law, for every thing that goes by the name of law. At the time of passing his law, let the legislator deliver, in the character of reasons, the considerations by which he was led to the passing of it [a].

This done, so long as in the eyes of the succeeding legislators the state of facts on which the reasons are grounded appears to continue without material change, and the reasons to appear satisfactory, so long the law continues: but no sooner do the reasons cease to appear satisfactory, or the state of the facts to have undergone any such change as to call for an alteration in the law, than an alteration in it, or the abrogation of it, takes place accordingly.

A declaration or assertion that this or that law is immutable, so far from being a proper instrument to ensure its permanency, is rather a presumption that such law has some mischievous tendency.

The better the law, the less is any such extraneous argument likely to be recurred to for the support of it; the worse the law, and thence the more completely destitute of all intrinsic support, the more likely is it that support should be sought for it from this extraneous source.

---

[a] For a specimen, see the end of the first volume of Dumont's *Traités de Législation.*

But though it is the characteristic tendency of this instrument to apply itself to bad laws in preference to good ones, there is another, the tendency of which is to apply itself to good ones in preference to bad : this is what may be termed justification; the practice of annexing to each law the considerations by which, in the character of *reasons*, the legislator was induced to adopt it [a]; a practice which, if rigidly pursued, must at no distant interval put an exclusion on all bad laws.

To the framing of laws so constituted, that, being good in themselves, an accompaniment of good and sufficient reasons should also be given for them, there would be requisite, in the legislator, a probity not to be diverted by the action of sinister interest, and intelligence adequate to an enlarged comprehension and close application of the principle of general utility: in other words, the principle of the greatest happiness of the greatest number.

But to draw up laws without reasons, and laws for which good reasons are not in the nature of the case to be found, requires no more than the union of will and power.

The man who should produce a body of good laws with an accompaniment of good reasons, would feel an honest pride at the prospect of holding thus in bondage a succession of willing generations; his triumph would be to leave them the power, but to

---

[a] See Bentham per Dumont *Traités de Législation,* &c.; *Papers on Codification;* and *Letters to the United States.*

deprive them of will to escape. But to the champions of abuse, by whom, amongst other devices, the conceit of immutable laws is played off against reform, in whatever shape it presents itself, every use of reason is as odious as the light of the sun to moles and burglars. ⨉

2. *Vows, or promissory Oaths.*

The object in this fallacy is the same as in the preceding; but to the absurdity involved in the notion of tying up the hands of generations yet to come is added, in this case, that which consists in the use sought to be made of supernatural power : the arm pressed into the service is that of the invisible and supreme ruler of the universe.

The oath taken, the formularies involved in it being pronounced, is or is not the Almighty bound to do what is expected of him? Of the two contradictory propositions, which is it that you believe?

If he is *not* bound, then the security, the sanction, the obligation amounts to nothing.

If he *is* bound, then observe the consequence :—the Almighty is bound; and by whom bound ?—of all the worms that crawl about the earth in the shape of men, there is not one who may not thus impose conditions on the supreme ruler of the universe.

And to what is he bound? to any number of contradictory and incompatible observances, which legislators, tyrants, or madmen, may, in the shape of an auth, be pleased to assign.

Eventual, it must be acknowledged, and no more, is the power thus exercised over, the task thus imposed upon, the Almighty. So long as the vow is kept, there is nothing for him to do ;—true : but no sooner is the vow broken, than his task commences ; a task which consists in the inflicting on him by whom the vow is broken a punishment, which, when it is inflicted, is of no use in the way of example, since nobody ever sees it.

The punishment, it may be said, when inflicted, will be such exactly as, in the judgment of the almighty and infallible judge, will be best adapted to the nature of the offence.

Yes: but what offence? not the act which the oath was intended to prevent, for that act may be indifferent, or even meritorious; and, if criminal, ought to be punished independently of the oath : the only offence peculiar to this case, is the profanation of a ceremony ; and the profanation is the same, whether the act by which the profanation arises be pernicious or beneficial.

It is in vain to urge, in this or that particular instance, in proof of the reasonableness of the oath, the reasonableness of the prohibition or command which it is thus employed to perpetuate.

The objection is to the principle itself: to any idea of employing an instrument so unfit to be employed.

No sort of security is given, or can be given, for the applying it to the most beneficial purpose rather than to the most pernicious.

On the contrary, it is more likely to be applied to a pernicious than to a beneficial purpose ;

Because, the more manifestly and undeniably beneficial the observance of the prohibition in question would be in the eyes of future generations, the more likely is the prohibition to be observed, independently of the oath : as, on the other hand, the more likely the prohibition is not to be observed otherwise, the greater is the demand for a security of this extraordinary complexion to enforce the observance.

We come now to the instance in which, by the operation of the fallacy here in question, the ceremony of an oath has been endeavoured to be applied to the perpetuation of misrule.

Among the statutes passed in the first parliament of William and Mary, is one entitled " An Act for establishing the Coronation Oath[a]."

The form in which the ceremony is performed is as follows :—By the archbishop or bishop, certain questions are put to the monarch ; and it is of the answers given to these questions that the oath is composed.

Of these questions, the third is as follows : " Will you, to the utmost of your power, maintain the laws of God, the true profession of the Gospel, and the protestant reformed religion established by law ? And will you preserve unto the bishops and clergy of this realm, and to the churches committed to their charge,

---

[a] 1 W. and M. c. 6. anno 1688.

all such rights and privileges as by law do or shall appertain unto them, or any of them?"

Answer. "All this I promise to do."

After this, anno 1706, comes the Act of Union, in the concluding article of which it is said, "That after the demise of her majesty . . . the sovereign next succeeding to her majesty in the royal government of the kingdom of Great Britain, and so for ever hereafter, every king or queen succeeding and coming to the royal government of the kingdom of Great Britain, at his or her coronation, shall in the presence," &c. "take and subscribe an oath to maintain and preserve inviolate the said settlement of the church, and the doctrine, worship, discipline and government thereof, as by law established, within the kingdoms of England and Ireland, the dominion of Wales, and town of Berwick-upon-Tweed, and the territories thereunto belonging[a]."

A notion was once started, and upon occasion may but too probably be broached again, that by the above clause in the coronation oath, the king stands precluded from joining in the putting the majority of the Irish upon an equal footing with the minority, as well as from affording to both together relief against the abuses of the ecclesiastical establishment of that country.

In relation to this notion, the following propositions have already, it is hoped, been put sufficiently out of doubt.

---

[a] 5 Ann. c 8. art. 25. § 8.

1. That it ought not to be in the power of the sovereignty to tie up its own hands, or the hands of its successors.

2. That, on the part of the sovereignty, no such power can have existence, either here or any where else.

3. That, therefore, all attempts to exercise any such power are, in their own nature, to use the technical language of lawyers, null and void.

4. Another, which will, it is supposed, appear scarcely less clear, is, that no such anarchical wish or expectation was entertained by the framers of the oath.

The proposition maintained is, that to any bills, to the effect in question, the monarch is, by this third and last clause in the oath, precluded from giving his assent : if so, he is equally precluded from giving his assent to any bills, to any proposed laws whatever.

It is plainly in what is called his executive, and not in his legislative capacity, that the obligation in question was meant to attach upon the monarch.

So loose are the words of the act, that, if they were deemed to apply to the monarch in his legislative capacity, he might find in them a pretence for refusing assent to almost any thing he did not like.

If by this third clause he stands precluded from consenting to any bill, the effect of which would be to abolish or vary any of the "rights" or "privileges" appertaining to the bishops or clergy, or "any of them," then by the first clause he stands equally precluded from giving his concurrence to any law, the effect of which would be to abolish or change any other rights. For by this first clause he is made "solemnly" to "promise and swear to govern the people . . . according to the statutes in parliament agreed on, and the laws and customs of the same." After this, governing according to any new law, he could not govern according to the old law abrogated by it.

If, by any such ceremony, misrule in this shape could be converted into a duty or a right, so might it in any other.

If Henry VIII. at his coronation had sworn to "maintain" that Catholic "religion," which for so many centuries was "established by law," and by fire and sword to keep out the Protestant religion, and had been considered bound by such oath, he could never have taken one step towards the Reformation, and the religion of the state must have been still Catholic.

But would you put a force upon the conscience of your sovereign ? By any construction, which in your judgment may be the proper one, would you preclude him from the free exercise of his ?

Most assuredly not: even were it as completely within as it is out of my power.

All I plead for is, that on so easy a condition as that of pronouncing the word *conscience*, it may not be in his power either to make himself absolute, or in any shape to give continuance to misrule.

Let him but resign his power, conscience can never reproach him with any misuse of it.

It seems difficult to say what can be a misuse of it, if it be not a determinate and persevering habit of using it in such a manner as in the judgment of the two houses is not " conducive," but repugnant " to the utility of the subjects," with reference to whom, and whose utility alone, either laws or kings can be of any use.

According to the form in which it is conceived, any such engagement is in effect either a check or a license :—a license under the appearance of a check, and for that very reason but the more efficiently operative.

Chains to the man in power? Yes :—but such as he figures with on the stage : to the spectators as imposing, to himself as light as possible. Modelled by the wearer to suit his own purposes, they serve to rattle but not to restrain.

Suppose a king of Great Britain and Ireland to have expressed his fixed determination, in the event of any proposed law being tendered to him for his assent, to refuse such assent, and this not on the persuasion that the law would not be " for the utility of the subjects," but that by his coronation oath he stands precluded from so doing :—the course proper to be taken by parliament, the course pointed out by principle and precedent would be, a vote of abdication : —a vote declaring the king to have abdicated his royal authority, and that, as in case of death or incurable mental derangement, now is the time for the person next in succession to take his place.

In the celebrated case in which a vote to this effect was actually passed, the declaration of abdication was in lawyers' language a fiction,—in plain truth a falsehood,—and that falsehood a mockery; not a particle of his power was it the wish of James to abdicate, to part with; but to increase it to a maximum was the manifest object of all his efforts.

But in the case here supposed, with respect to a part, and that a principal part of the royal authority, the will and purpose to abdicate is actually declared: and this, being such a part, without which the remainder cannot, "to the utility of the subjects," be exercised, the remainder must of necessity be, on their part and for their sake, added [a].

---

[a] The variety of the notions entertained at different periods, in different stages of society, respecting the duration of laws, presents a curious and not uninstructive picture of human weakness.

1. At one time we see, under the name of king, a single person, whose will makes law, or, at any rate, without whose will no law is made; and when this law-giver dies, his laws die with him.

Such was the state of things in Saxon times,—such even continued to be the state of things for several reigns after the Norman Conquest [*].

2. Next to this comes a period in which the duration of the law, during the life-time of the monarch to whom it owed its birth, was unsettled and left to chance [†].

3. In the third place comes the period in which the notions respecting the duration of the law concur with the dictates of reason and utility, not so much from reflection as because no occasion of a

---

[*] To Ric. I. inclusive.
[†] John, Ed. I and II.

nature to suggest and urge any attempt so absurd as that of tyrannizing over futurity had as yet happened to present itself.

4. Lastly, upon the spur of an occasion of the sort in question, comes the attempt to give eternity to human laws.

Provisional and eventual perpetuity is an attribute which, in that stage of society at which laws have ceased to expire with the individual legislator, is understood to be inherent in all laws in which no expression is found to the contrary.

But if a particular length of time be marked out, during which, in the enactment of a law, it is declared that that law shall not be liable to suffer abrogation or alteration, the determination to tie up the hands of succeeding legislators is expressed in unequivocal terms.

Such, in respect of their constitutional code, was the pretension set up by the first assembly of legislators brought together by the French revolution.

A position not less absurd in principle, but by the limitation in point of time, not pregnant with any thing like equal mischief, was before that time acted upon, and still continues to be acted upon, in English legislation.

In various statutes, a clause may be found by which the statute is declared capable of being altered or repealed in the course of the same session. In this clause is contained, in the way of necessary implication, that a statute in which no such clause is inserted is not capable of being repealed or altered during the session,—no, not by the very hands by which it was made.

CHAPTER IV.

No-precedent Argument.

Ad verecundiam.

Exposition.

"THE proposition is of a novel and unprecedented complexion : the present is surely the first time that any such thing was ever heard of in this house."

Whatsoever may happen to be the *subject* introduced, above is a specimen of the infinite variety of forms in which the opposing *predicate* may be clothed.

To such an observation there could be no objection, if the object with which it were made was only to fix attention to a new or difficult subject : " Deliberate well before you act, as you have no precedent to direct your course :"

Exposure.

But in the character of an argument, as a ground for the rejection of the proposed measure, it is obviously a fallacy.

Whether or no the alleged novelty actually exists, is an inquiry which it can never be worth while to make.

That it is impossible that it should in any case afford the smallest ground for the rejection of the measure,—that the observation is completely irrelevant in

I

relation to the question, whether or no it is expedient that such a measure should be adopted,—is a proposition to which it seems difficult to conceive how an immediate assent can be refused. If no specific good is indicated as likely to be produced by the proposed measure, this deficiency is itself sufficient to warrant the rejection of it. If any such specific good *is* indicated, it must be minute indeed if an observation of this nature can afford a sufficient ground for the rejection of the measure.

If the observation presents a conclusive objection against the particular measure proposed, so it would against any other that ever was proposed, including every measure that ever was adopted, and therein every institution that exists at present. If it proves that this ought not to be done, it proves that nothing else ought ever to have been done.

It may be urged, that, if the measure had been a fit one, it would have been brought upon the carpet before. But there are several obstacles besides the inexpediency of a measure, which, for any length of time, may prevent its being brought forward.

1. If, though beyond dispute promotive of the interest of the many, there be any thing in it that is adverse to the interests, the prejudices, or the humours of the ruling few, the wonder is, not that it should not have been brought forward before, but that it should be brought forward even now.

2. If, in the complexion of it, there be any thing which it required a particular degree of ingenuity to

contrive and adapt to the purpose, this would of itself be sufficient to account for the tardiness of its appearance.

In legislation, the birth of ingenuity is obstructed and retarded by difficulties, beyond any which exist in other matters. Besides the more general sinister interest of the powerful few in whose hands the functions of government are lodged, the more particular sinister interest affecting the body of lawyers, is one to which any given measure, in proportion to the ingenuity displayed in it, is likely to be adverse.

Measures which come under the head of indirect legislation, and in particular those which have the quality of executing themselves, are the measures which, as they possess most efficiency when established, so they require greater ingenuity in the contrivance. Now in proportion as laws execute themselves, in other words, are attended with voluntary obedience, in that proportion are they efficient; but it is only in proportion as they fail of being efficient, that, to the man of law, they are beneficial and productive; because it is only in proportion as they stand in need of enforcement, that business makes its way into the hands of the man of law.

CHAPTER V.

Self-assumed Authority.

Ad ignorantiam; ad verecundiam.

THIS fallacy presents itself in two shapes:—1. An avowal made with a sort of mock modesty and caution by a person in exalted station, that he is incapable of forming a judgment on the question in debate, such incapacity being sometimes real, sometimes pretended : 2. Open assertion by a person so situated of the purity of his motives and integrity of his life, and the entire reliance which may consequently be reposed on all he says or does.

Sect. 1.

The first is commonly played off as follows:—An evil or defect in our institutions is pointed out clearly, and a remedy proposed, to which no objection can be made;—up starts a man high in office, and, instead of stating any specific objection, says, " I am not prepared " to do so and so, " I am not prepared to say," &c. The meaning evidently intended to be conveyed is, " If I, who am so dignified and supposed to be so capable of forming a judgment, avow myself incompetent to do so, what presumption, what folly must there be in the conclusion formed by any one else!" In truth, this is nothing else but an indirect way of browbeating ;—arrogance under a thin veil of modesty.

If you are not prepared to pass a judgment, you

are not prepared to condemn, and ought not, therefore, to oppose: the utmost you are warranted in doing, if sincere, is to ask for a little time for consideration.

Supposing the unpreparedness real, the reasonable and practical inference is,—say nothing, take no part in the business.

A proposition for the reforming of this or that abuse in the administration of justice, is the common occasion for the employment of this fallacy.

In virtue of his office, every judge, every law-officer, is supposed and pronounced to be profoundly versed in the science of the law;

Yes; of the science of the law as it is, probably as much as any other man : but law, as it ought to be, is a very different thing; and the proposal in question has for its avowed, and commonly for its real object, the bringing law as it is somewhat nearer to law as it ought to be. But this is one of those things for which the great dignitary is sure to be at all times unprepared :—unprepared to join in any such design ; every thing of this sort having been at all times contrary to his interest :—unprepared so much as to form any judgment concerning the conduciveness of the proposed measure to such its declared object: in any such point of view it has never been his interest to consider it.

A mind that, from its first entrance upon this subject, has been applying its whole force to the inquiry, as to what are the most effectual means of making its profit of the imperfections of the system ;—a mind to which of consequence the profit from these sources of affliction has been all along an object of complacency, and the affliction itself, at best, but an object of indifference ;—a mind which has, throughout the whole course of its career, been receiving a correspondent bias, and has in consequence contracted a correspondent distortion ;—cannot with reason be expected to exert itself with much alacrity or facility in a track so opposite and so new.

For the quiet of his conscience, if, at the outset of his career, it were his fortune to have one, he will naturally have been feeding himself with the notion, that, if there be any thing that is amiss, in practice it cannot be otherwise ; which being granted, and, accordingly, that suffering to a certain amount cannot but take place, whatsoever profit can be extracted from it, is fair game, and as such, belongs of right to the first occupant among persons duly qualified.

The wonder would not be great if an officer of the military profession should exhibit, for a time at least, some awkwardness if forced to act in the character of a surgeon's mate : to inflict wounds requires one sort of skill, to dress and heal them requires another. Telephus is the only man upon record who possessed an instrument by which wounds were with equal dispatch and efficiency made and healed. The race of Telephus is extinct ; and as to his spears, if ever any of them found their way into Pompeii or Herculaneum, they remain still among the ruins.

Unfortunately in this case, were the ability to form a judgment ever so complete, the likelihood of co-operation would not be increased. None are so completely deaf as those who will not hear,—none are so completely unintelligent as those who will not understand.

Call upon a chief justice to concur in a measure for giving possibility to the recovery of a debt, the recovery of which is in his own court rendered impossible by costs which partly go into his own pocket, as well might you call upon the Pope to abjure the errors of the church of Rome. If not hard pressed, he will maintain a prudent and easy silence ; if hard pressed, he will let fly a volley of fallacies : he will play off the argument drawn from the imputation of bad motives, and tell you of the profit expected by the party by whom the bill was framed, and petition procured, to form a ground for it. If that be not sufficient, he will transform himself in the first place into a witness, giving evidence upon a committee; in the next place, after multiplying himself into the number of members necessary to hear and report upon that evidence, he will make a report accordingly.

He will report in that character, that when in any town a set of tradesmen have, on their petition, obtained a judicatory in which the recovery of a debt under 40*s.* or 5*l.* is not attended with that obstruction of accumulated expense by which the relief which his judicatory professes to afford is always accompanied,

it has been with no other effect than that of giving in
the character of judges effect to claims, which in the
character of witnesses it was originally their design,
and afterwards their practice, to give support to by
perjury.

Sect. 2. *The second of these two devices may be called*
The Self-trumpeter's Fallacy.

By this name it is not intended to designate those
occasional impulses of vanity which lead a man to dis-
play or overrate his pretensions to superior intelligence.   Against the self-love of the man whose altar
to himself is raised on this ground, rival altars, from
every one of which he is sure of discouragement,
raise themselves all around.

But there are certain men in office who, in discharge
of their functions, arrogate to themselves a degree of
probity, which is to exclude all imputations and all
inquiry : their assertions are to be deemed equivalent
to proof ; their virtues are guarantees for the faithful
discharge of their duties ; and the most implicit confidence is to be reposed in them on all occasions.   If
you expose any abuse, propose any reform, call for
securities, inquiry, or measures to promote publicity,
they set up a cry of surprise, amounting almost to indignation, as if their integrity were questioned, or their
honour wounded.   With all this, they dexterously
mix up intimations, that the most exalted patriotism,
honour, and perhaps religion, are the only sources of
all their actions.

Such assertions must be classed among fallacies, because, 1. they are irrelevant to the subject in discussion : 2. the degree in which the predominance of motives of the social or disinterested cast is commonly asserted or insinuated, is, by the very nature of man, rendered impossible : 3. the sort of testimony thus given affords no legitimate reason for regarding the assertion in question to be true ; for it is no less completely in the power of the most profligate than in that of the most virtuous of mankind : nor is it in a less degree the interest of the profligate man to make such assertions. Be they ever so completely false, not any the least danger of punishment does he see himself exposed to, at the hands either of the law or of public opinion.

For ascribing to any one of these self-trumpeters the smallest possible particle of that virtue which they are so loud in the profession of, there is no more rational cause, than for looking upon this or that actor as a good man because he acts well the part of Othello, or bad because he acts well the part of Iago.

4. On the contrary, the interest he has in trying what may be done by these means, is more decided and exclusive than in the case of the man of real probity and social feeling. The virtuous man, being what he is, has that chance for being looked upon as such ; whereas the self-trumpeter in question, having no such ground of reliance, beholds his only chance in the conjunct effect of his own effrontery, and the imbecility of his hearers.

These assertions of authority, therefore, by men in office, who would have us estimate their conduct by their character, and not their character by their conduct, must be classed among political fallacies. If there be any one maxim in politics more certain than another, it is, that no possible degree of virtue in the governor can render it expedient for the governed to dispense with good laws and good institutions [a].

---

[a] Madame de Stael says, that in a conversation which she had at Petersburgh with the Emperor of Russia, he expressed his desire to better the condition of the peasantry, who are still in a state of absolute slavery; upon which the female sentimentalist exclaimed, " Sire, your character is a constitution for your country, and your conscience is its guarantee." His reply was, " *Quand cela serait, je ne serais jamais* qu'un accident heureux."—*Dix Années d'Exil,* p. 313.

CHAPTER VI.

Laudatory Personalities.

Ad amicitiam.

PERSONALITIES of this class are the opposites, and in some respects the counterparts, of vituperative personalities, which will be treated of next in order, at the commencement of the ensuing book.

Laudatory personalities are susceptible of the same number of modifications as will be shown to exist in the case of vituperative personalities : but in this case the argument is so much weaker than in the other, that the shades and modifications of it are seldom resorted to, and are therefore not worth a detailed exposition. The object of vituperative personalities is to effect the rejection of a measure, on account of the alleged bad character of those who promote it ; and the argument advanced is, "The persons who propose or promote the measure, are bad : therefore the measure is bad, or ought to be rejected." The object of laudatory personalities is to effect the rejection of a measure on account of the alleged good character of those who oppose it ; and the argument advanced is, "The measure is rendered unnecessary by the virtues of those who are in power,—their opposition is a sufficient authority for the rejection of the measure."

The argument indeed is generally confined to persons of this description, and is little else than an extension of the self-trumpeter's fallacy. In both of them, authority derived from the virtues or talents of the persons lauded is brought forward as superseding the necessity of all investigation.

"The measure proposed implies a distrust of the members of His Majesty's Government; but so great is their integrity, so complete their disinterestedness, so uniformly do they prefer the public advantage to their own, that such a measure is altogether unnecessary. Their disapproval is sufficient to warrant an opposition; precautions can only be requisite where danger is apprehended; here, the high character of the individuals in question is a sufficient guarantee against any ground of alarm."

The panegyric goes on increasing in proportion to the dignity of the functionary thus panegyrized.

Subordinates in office are the very models of assiduity, attention, and fidelity to their trust; ministers, the perfection of probity and intelligence : and as for the highest magistrate in the state, no adulation is equal to describe the extent of his various merits.

There can be no difficulty in exposing the fallacy of the argument attempted to be deduced from these panegyrics.

1st, They have the common character of being irrelevant to the question under discussion. The measure must have something extraordinary in it, if a right judgment cannot be founded on its merits without first estimating the character of the members of the Government.

2nd, If the goodness of the measure be sufficiently established by direct arguments, the reception given to it by those who oppose it, will form a better criterion for judging of their character, than their character, (as inferred from the places which they occupy,) for judging of the goodness or badness of the measure.

3rd, If this argument be good in any one case, it is equally good in every other ; and the effect of it, if admitted, would be to give to the persons occupying for the time being the situation in question, an absolute and universal negative upon every measure not agreeable to their inclinations.

4th, In every public trust, the legislator should, for the purpose of prevention, suppose the trustee disposed to break the trust in every imaginable way in which it would be possible for him to reap, from the breach of it, any personal advantage. This is the principle on which public institutions ought to be formed ; and when it is applied to all men indiscriminately, it is injurious to none. The practical inference is, to oppose to such possible (and what will always be probable) breaches of trust every bar that can be opposed, consistently with the power requisite for the efficient and due discharge of the trust. Indeed, these arguments, drawn from the supposed virtues of men in power, are opposed to the first principles on which all laws proceed.

5th, Such allegations of individual virtue are never

supported by specific proof, are scarce ever susceptible of specific disproof; and specific disproof, if offered, could not be admitted : viz. in either house of
parliament. If attempted elsewhere, the punishment
would fall, not on the unworthy trustee, but on him
by whom the unworthiness had been proved.

PART THE SECOND.

FALLACIES OF DANGER,

The subject matter of which is Danger in various shapes, and the object, to repress discussion altogether, by exciting alarm.

CHAPTER I.

Vituperative Personalities.

Ad odium.

To this class belongs a cluster of fallacies so intimately connected with each other, that they may first be enumerated and some observations be made upon them in the lump. By seeing their mutual relations to each other, by observing in what circumstances they agree, and in what they differ, a much more correct as well as complete view will be obtained of them, than if they were considered each of them by itself.

The fallacies that belong to this cluster may be denominated,

1. Imputation of bad design.
2. Imputation of bad character.
3. Imputation of bad motive.

4. Imputation of inconsistency.

5. Imputation of suspicious connexions.—*Noscitur ex sociis.*

6. Imputation founded on identity of denomination.—*Noscitur ex cognominibus.*

Of the fallacies belonging to this class, the common character is the endeavour to draw aside attention from the *measure* to the *man* [a]; and this in such sort as, from the supposed imperfection on the part of the man by whom a measure is supported or opposed, to cause a correspondent imperfection to be imputed to the measure so supported, or excellence to the measure so opposed. The argument in its various shapes

---

[a] On the subject of personalities of the vituperative kind, the following are the instructions given by Gerard Hamilton: they contain all he says upon the subject. I. 31. 367. p. 67. "It is an artifice to be used (but if used by others, to be detected), to begin some personality, or to throw in something that may bring on a personal altercation, and draw off the attention of the House from the main point." II. 36. (470) p. 86. "If your cause is too bad, call, in aid, the party " (meaning, probably, the *individual* who stands in the situation of party, not the assemblage of men of whom a political party is composed): " if the party is bad, call, in aid, the cause : if neither is good, *wound the opponent.*" III. "If a person is powerful, he is to be made obnoxious ; if helpless, contemptible : if wicked, detestable." In this we have, so far as concerns the head of personalities, " the whole fruit and result of the experience of one who was by no means unconversant with law " (says his editor, p. 6), " and had himself sat in Parliament for more than forty years ; . . . devoting almost all his leisure and thoughts, during the long period above mentioned, to the examination and discussion of all the principal questions agitated in Parliament, and of the several topics and modes of reasoning by which they were either supported or opposed."

amounts to this :—In bringing forward or supporting
the measure in question, the person in question entertains a bad design ; therefore the measure is bad :—
he is a person of a bad character, therefore the measure is bad :—he is actuated by a bad motive, therefore the measure is bad :—he has fallen into incon-
sistencies ; on a former occasion he either opposed it,
or made some observation not reconcileable with some
observation which he has advanced on the present
occasion ; therefore the measure is bad :—he is on a
footing of intimacy with this or that person, who is a
man of dangerous principles and designs, or has been
seen more or less frequently in his company, or has
professed or is suspected of entertaining some opinion
which the other has professed, or been suspected of
entertaining; therefore the measure is bad :—he bears
a name that at a former period was borne by a set of
men now no more, by whom bad principles were entertained, or bad things done ; therefore the measure
is bad.

In these arguments thus arranged, a sort of anti-
climax may be observed ; the fact intimated by each
succeeding argument being suggested in the character
of evidence of the one immediately preceding it, or at
least of some one or more of those which precede it,
and the conclusion being accordingly weaker and
weaker at each step. The second is a sort of circumstantial evidence of the first, the third of the second,
and so on. If the first is inconclusive, the rest fall at
once to the ground.

K

Exposure.

Various are the considerations which concur in demonstrating the futility of the fallacies comprehended in this class, and (not to speak of the improbity of the utterers) the weakness of those with whom they obtain currency,—the weakness of the acceptors.

1. In the first place, comes that general character of irrelevancy which belongs to these, in common with the several other articles that stand upon the list of fallacies.

2. In the next place, comes the complete inconclusiveness. Whatsoever be their force as applied to a bad measure, to the worst measure that can be imagined, they would be found to apply with little less force to all good measures, to the best measures that can be imagined.

Among 658 or any such large number of persons taken at random, there will be persons of all characters : if the measure is a good one, will it become bad because it is supported by a bad man ? If it is bad, will it become good because supported by a good man ? If the measure be really inexpedient, why not at once show that it is so?—Your producing these irrelevant and inconclusive arguments in lieu of direct ones, though not sufficient to prove that the measure you thus oppose is a good one, *contributes* to prove that you yourselves regard it as a good one.

After these general observations, let us examine, more in detail, the various shapes the fallacy assumes.

Sect. 1. To begin with the *Imputation of bad design.*

The measure in question is not charged with being
itself a bad one; for if it be, and in so far as it is thus
charged, the argument is not irrelevant and fallacious.
The bad design imputed, consists not in the design of
carrying this measure, but some other measure, which
is thus, by necessary implication, charged with being
a bad one. Here, then, four things ought to be proved:
viz.—1. That the design of bringing forward the supposed bad measure is really entertained : 2. That this
design will be carried into effect: 3. That the measure
will prove to be a bad one : 4. That, but for the actually proposed measure, the supposed bad one would
not be carried into effect.

This is, in effect, a modification of *the fallacy of
distrust,* which will shortly be treated of.

But on what ground rests the supposition, that the
supposed bad measure will, as such a consequence,
be carried into effect ?   The persons by whom, if at
all, it will be carried into effect, will be, either the
legislators for the time being, or the legislators of
some future contingent time : as to the legislators for
the time being, observe the character and frame of
mind which the orator imputes to these his judges ;—
" Give not your sanction to this measure; for though
there may be no particular harm in it, yet, if you do
give your sanction to it, the same man by whom this
is proposed, will propose to you others that will be
bad ; and such is your weakness, that, however bad

they may be, you will want either the discernment
necessary to enable you to see them in their true light,
or the resolution to enable you to put a negative upon
measures, of the mischief of which you are fully convinced." The imbecility of the persons thus addressed
in the character of legislators and judges, their consequent unfitness for the situation,—such, it is manifest,
is the basis of this fallacy. On the part of these legislators themselves, the forbearance manifested under
such treatment,—on the part of the orator, the confidence entertained of his experiencing such forbearance,
—afford no inconsiderable presumption of the reality
of the character so imputed to them.

Sect. 2. *Imputation of bad character.*

The inference meant to be drawn from an imputation of bad character is, either to cause the person in
question to be considered as entertaining bad design,
i. e. about to be concerned in bringing forward future
contingent and pernicious measures, or simply to destroy any persuasive force, with which, in the character of authority, his opinion is likely to be attended.

In this last case, it is a fallacy opposed to a fallacy
of the same complexion, played off on the other side :
to employ it, is to combat the antagonist with his own
weapons. In the former case, it is another modification of *the fallacy of distrust*, of which, hereafter.

In proportion to the degree of efficiency with which
a man suffers these instruments of deception to operate
upon his mind, he enables bad men to exercise over

him a sort of power, the thought of which ought to cover him with shame. Allow this argument the effect of a conclusive one, you put it into the power of any man to draw you at pleasure from the support of every measure, which in your own eyes is good, to force you to give your support to any and every measure which in your own eyes is bad. Is it good?—the bad man embraces it, and, by the supposition, you reject it. Is it bad?—he vituperates it, and that suffices for driving you into its embrace. You split upon the rocks, because he has avoided them; you miss the harbour, because he has steered into it.

Give yourself up to any such blind antipathy, you are no less in the power of your adversaries than by a correspondently irrational sympathy and obsequiousness you put yourself into the power of your friends.

Sect. 3. *Imputation of bad motive.*

The proposer of the measure, it is asserted, is actuated by bad motives, from whence it is inferred that he entertains some bad design. This, again, is no more than a modification of *the fallacy of distrust;* but one of the very weakest; 1. because motives are hidden in the human breast; 2. because, if the measure is beneficial, it would be absurd to reject it on account of the motives of its author. But what is peculiar to this particular fallacy, is the falsity of the supposition on which it is grounded: viz. the existence of a class or species of motives, to which any such epithet as bad can, with propriety, be applied. What constitutes a

motive, is the eventual expectation, either of some
pleasure, or exemption from pain ; but forasmuch as
in itself there is nothing good but pleasure, or exemption from pain, it follows that no motive is bad in itself, though every kind of motive may, according to
circumstances, occasion good or bad actions [a] ; and
motives of the dissocial cast may aggravate the mischief of a pernicious act.   But if the act itself to
which the motive gives birth,—if, in the proposed
measure in question, there be nothing pernicious,—
it is not in the motive's being of the dissocial class,—
it is not in its being of the self-regarding class,—that
there is any reason for calling it a bad one.   Upon
the influence and prevalence of motives of the self-
regarding class, depends the preservation, not only of
the species, but of each individual belonging to it.
When, from the introduction of a measure, a man
beholds the prospect of personal advantage in any
shape whatever to himself,—say, for example, a pecuniary advantage, as being the most ordinary and
palpable, or, dyslogistically speaking, the most gross,
—it is certain that the contemplation of this advantage must have had some share in causing the conduct
he pursues—it may have been the only cause.   The
measure itself being by the supposition not pernicious,
is it the worse for this advantage?   On the contrary, it
is so much the better.   For of what stuff is public advantage composed, but of private and personal advantage?

---

[a] See Dumont *Traités de Législation,* tom. ii. c. 8. ed. 2 ; Bentham,
Theory of Morals and Legislation.

Sect. 4. *Imputation of inconsistency.*

Admitting the fact of the inconsistency, the utmost it can amount to in the character of an argument against the proposed measure, is, the affording a presumption of bad design in a certain way, or of bad character in a certain way and to a certain degree, on the part of the proposer or supporter of the measure. Of the futility of that argument, a view has been already given; and this, again, is a modification of *the fallacy of distrust*.

That inconsistency, when pushed to a certain degree, may afford but too conclusive evidence of a sort of relatively bad character, is not to be denied : if, for example, on a former occasion, personal interest inclining him one way (say against the measure), arguments have been urged by the person in question against the measure, while on the present occasion personal interest inclining him the opposite way, arguments are urged by him in favour of the measure,— or if a matter of fact, which on a former occasion was denied, be now asserted, or *vice versâ*,—and in each case if no notice of the inconsistency is taken by the person himself,—the operation of it to his prejudice will naturally be stronger than if an account more or less satisfactory is given by him of the circumstances and causes of the variance.

But be the evidence with regard to the cause of the change what it may, no inference can be drawn from it against the measure unless it be that such inconsistency, if established, may weaken the persuasive force of the opinion of the person in question in the character of authority : and in what respect and degree an argument of this complexion is irrelevant, has been already brought to view.

Sect. 5. *Imputation of suspicious connexions.*
Noscitur ex sociis.

The alleged badness of character, on the part of the alleged associate, being admitted, the argument now in question will stand upon the same footing as the four preceding; the weakness of which has been already exposed, and will constitute only another branch of *the fallacy of distrust*. But before it can stand on a par even with those weak ones, two ulterior points remain to be established.

1. One, is the badness of character on the part of the alleged associate.

2. Another, is the existence of a social connexion between the person in question and his supposed associate.

3. A third, is, that the influence exercised on the mind of the person in question is such, that in consequence of the connexion he will be induced to introduce and support measures (and those mischievous ones), which otherwise he would not have introduced or supported.

As to the two first of these three supposed facts, their respective degrees of probability will depend on the circumstances of each case. Of the third, the

weakness may be exposed by considerations of a general nature.   In private life, the force of the presumption in question is established by daily experience : but in the case of a political connexion, such
as that which is created by an opposition to one and
the same political measure or set of measures, the
presumption loses a great part, sometimes the whole,
of its force.   Few are the political measures, on the
occasion of which men of all characters, men of all
degrees, in the scale of probity and improbity, may
not be seen on both sides.

The mere need of information respecting matters
of fact, is a cause capable of bringing together, in a
state of apparent connexion, some of the most opposite characters.

Sect. 6. *Imputation founded on identity of denomination.*
Noscitur ex cognominibus.

The circumstances by which this fallacy is distinguished from the last preceding, is, that in this case
between the person in question and the obnoxious persons by whose opinions and conduct he is supposed
to be determined or influenced, neither personal intercourse nor possibility of personal intercourse can exist.
In the last case, his measures were to be opposed because he was connected with persons of bad character; in the present, because he bears the same denomination as persons now no more, but who, in their
own time, were the authors of pernicious measures.
In so far as a community of interest exists between

the persons thus connected by community of denomination, the allegation of a certain community of designs is not altogether destitute of weight.   Community of denomination, however, is but the sign, not the
efficient cause, of community of interest.   What have
the Romans of the present day in common with the
Romans of early times ?   Do they aspire to recover
the empire of the world ?

But when evil designs are imputed to men of the
present day, on the ground that evil designs were entertained and prosecuted by their namesakes in time
past, whatsoever may be the community of interest,
one circumstance ought never to be out of mind :—
this is, the gradual melioration of character from the
most remote and barbarous, down to the present
time; the consequence of which is, that in many par-
ticulars the same ends which were formerly pursued
by persons of the same denomination are not now
pursued ; and if in many others the same ends are
pursued, they are not pursued by the same bad means.
If this observation pass unheeded, the consequences
may be no less mischievous than absurd : that which
has been, is unalterable.   If, then, this fallacy be
suffered to influence the mind and determine human
conduct, whatsoever degree of depravity be imputed
to preceding generations of the obnoxious denomination,—whatsoever opposition may have been manifested towards them or their successors,—must continue without abatement to the end of time. " Be my
friendship immortal, my enmity mortal," is the senti-

ment that has been so warmly and so justly applaud-
ed in the mouth of a sage of antiquity : but the fallacy
here in question proposes to maintain its baneful influence for ever.

It is in matters touching religious persuasion, and
to the prejudice of certain sects, that this fallacy has
been played off with the greatest and most pernicious
effect. In England, particularly against measures for
the relief of the Catholics, " those of our ancestors,
who, professing the same branch of the Christian religion as that which you now profess, were thence distinguished by the same name, entertained pernicious
designs, that for some time showed themselves in pernicious measures; therefore you, entertaining the same
pernicious designs, would now, had you but power
enough, carry into effect the same pernicious measures :—they, having the power, destroyed by fire and
faggot those who, in respect of religious opinions and
ceremonies, differed from them ; therefore, had you
but power enough, so would you." Upon this ground,
in one of the three kingdoms, a system of government
continues, which does not so much as profess to have
in view the welfare of the majority of the inhabitants,
—a system of government in which the interest of the
many is avowedly, so long as the government lasts,
intended to be kept in a state of perpetual sacrifice to
the interest of the few. In vain is it urged, these inferences, drawn from times and measures long since
past, are completely belied by the universal experience
of all present time. In the Saxon kingdom, in the

Austrian empire, in the vast and ever-flourishing empire of France, though the sovereign is Catholic, whatsoever degree of security the Government allows of is possessed alike by Catholics and Protestants. In vain is it observed (not that to this purpose this or any other part of the history of the 17th century is worth observing), in vain is it observed, and truly observed, the church of England continued her fires after the church of Rome had discontinued hers[n];

It is only in the absence of interest that experience can hope to be regarded, or reason heard. In the character of sinecurists and over-paid placemen, it is the interest of the members of the English Government to treat the majority of the people of Ireland on the double footing of enemies and subjects; and such is the treatment which is in store for them to the extent of their endurance.

Sect. 7. *Cause of the prevalence of the fallacies belonging to this class.*

Whatsoever be the nature of the several instruments of deception by which the mind is liable to be operated upon and deceived,—the degree of prevalence they experience,—the degree of success they enjoy,—depends ultimately upon one common cause: viz. the ignorance and mental imbecility of those on whom they operate. In the present instance,

---

[n] Under James I., when, for being Anabaptists or Arians, two men were burnt in Smithfield.

besides this ultimate cause or root, they find in another fallacy, and the corresponding propensity of the human mind, a sort of intermediate cause. This is the fallacy of authority: the corresponding propensity is the propensity to save exertion by resting satisfied with authority. Derived from, and proportioned to, the ignorance and weakness of the minds to which political arguments are addressed, is the propensity to judge of the propriety or impropriety of a measure from the supposed character or disposition of its supporters or opposers, in preference to, or even in exclusion of, its own intrinsic character and tendency. Proportioned to the degree of importance attached to the character and disposition of the author or supporter of the measure, is the degree of persuasive force with which the fallacies belonging to this class will naturally act.

Besides, nothing but laborious application, and a clear and comprehensive intellect, can enable a man on any given subject to employ successfully relevant arguments drawn from the subject itself. To employ personalities, neither labour nor intellect is required : in this sort of contest, the most idle and the most ignorant are quite on a par with, if not superior to, the most industrious and the most highly-gifted individuals. Nothing can be more convenient for those who would speak without the trouble of thinking; the same ideas are brought forward over and over again, and all that is required is to vary the turn of expression. Close and relevant arguments have very little hold on the passions, and serve rather to quell than to inflame them; while in personalities, there is always something stimulant, whether on the part of him who praises or him who blames. Praise forms a kind of connexion between the party praising and the party praised, and vituperation gives an air of courage and independence to the party who blames.

Ignorance and indolence, friendship and enmity, concurring and conflicting interest, servility and independence, all conspire to give personalities the ascendancy they so unhappily maintain. The more we lie under the influence of our own passions, the more we rely on others being affected in a similar degree. A man who can repel these injuries with dignity may often convert them into triumph: " Strike me, but hear," says he; and the fury of his antagonist redounds to his own discomfiture.

CHAPTER II.

The Hobgoblin Argument, or, No Innovation!

Ad metum.

Exposition.

THE hobgoblin, the eventual appearance of which is denounced by this argument, is *Anarchy ;* which tremendous spectre has for its forerunner the monster *Innovation.* The forms in which this monster may be denounced are as numerous and various as the sentences in which the word *innovation* can be placed.

" *Here it comes !* " exclaims the barbarous or unthinking servant in the hearing of the affrighted child, when, to rid herself of the burthen of attendance, such servant scruples not to employ an instrument of terror, the effects of which may continue during life. " *Here it comes !* " is the cry ; and the hobgoblin is rendered but the more terrific by the suppression of its name.

Of a similar nature, and productive of similar effects, is the political device here exposed to view. As an instrument of deception, the device is generally accompanied by personalities of the vituperative kind. Imputation of bad motives, bad designs, bad conduct and character, &c. are ordinarily cast on the authors and advocates of the obnoxious measure ; whilst the term employed is such as to beg the question in dispute. Thus, in the present instance, *inno-*

vation means a *bad* change, presenting to the mind, besides the idea of a *change*, the proposition, either that change in general is a bad thing, or at least that the sort of change in question is a bad change.

Exposure.

All-comprehensiveness of the condemnation passed by this fallacy.

This is one of the many cases in which it is difficult to render the absurdity of the argument more glaring than it is upon the face of the argument itself.

Whatever reason it affords for looking upon the proposed measure, be it what it may, as about to be mischievous, it affords the same reason for entertaining the same opinion of every thing that exists at present. To say all new things are bad, is as much as to say all things are bad, or, at any event, at their commencement : for of all the old things ever seen or heard of, there is not one that was not once new. Whatever is now *establishment* was once *innovation*.

He who on this ground condemns a proposed measure, condemns, in the same breath, whatsoever he would be most averse to be thought to disapprove.— He condemns the Revolution, the Reformation, the assumption made by the House of Commons of a part in the penning of the laws in the reign of Henry VI., the institution of the House of Commons itself in the reign of Henry III.,—all these he bids us regard as sure forerunners of the monster Anarchy, but particularly the birth and first efficient agency of the House of Commons; an innovation, in comparison of which all

others, past or future, are for efficiency, and consequently mischievousness, but as grains of dust in the
balance.

Sect. 2. *Apprehension of mischief from change, what
foundation it has in truth.*

A circumstance that gives a sort of colour to the
use of this fallacy is, that it can scarcely ever be found
without a certain degree of truth adhering to it. Supposing the change to be one which cannot be effected
without the interposition of the legislature, even this
circumstance is sufficient to attach to it a certain
quantity of mischief. The words necessary to commit the change even to writing cannot be put into
that form without labour, importing a proportional
quantity of vexation to the head employed in it; which
labour and vexation, if paid for, is compensated by
and productive of expense. When disseminated by
the operation of the press, as it always must be, before
it can be productive of whatever effect is aimed at, it
becomes productive of ulterior vexation and expense.
Here, then, is so much unavoidable mischief, of which
the most salutary and indispensable change cannot
fail to be productive: to this natural and unavoidable
portion of mischief, the additions that have been made
in the shape of factitious and avoidable mischief of the
same kind are such as have sufficient claim to notice,
but to a notice not proper for this place.

Here, then, we have the *minimum* of mischief,

L

which accompanies every change ; and in this minimum of mischief we have the minimum of truth with
which this fallacy is accompanied, and which is sufficient to protect it against exposure, from a flat and
undiscriminating denial.

It is seldom, however, that the whole of the mischief, with the corresponding portion of truth, is confined within such narrow bounds.

Wheresoever any portion, however great or small,
of the aggregate mass of the objects of desire in any
shape,—matter of wealth, power, dignity, or even reputation ;—and whether in possession or only in prospect, and that ever so remote and contingent, must,
in consequence of the change, pass out of any hand
or hands that are not willing to part with it,—viz.
either without compensation, or with no other than
what, in their estimation, is insufficient,—here we
have, in some shape or other, a quantity of vexation
uncompensated: so much vexation, so much mischief
beyond dispute.

But in one way or other, whether from the total
omission of this or that item, or from the supposed
inadequacy of the compensation given for it, or from
its incapacity of being included in any estimate, as in
case of remote and but weakly probable as well as
contingent profits, it will not unfrequently happen that
the compensation allotted in this case shall be inadequate, not only to the desires, but to the imagined
rights of the party from whom the sacrifice is exacted.

In so far as such insufficiency appears to himself to exist, he will feel himself urged by a motive, the force of which will be in proportion to the amount of such deficiency, to oppose the measure : and in so far as in his eyes such motive is fit to be displayed, it will constitute what in his language will be *reason*, and what will be received in that character by all other persons in whose estimate any such deficiency shall appear to exist. So far as any such deficiency is specifically alleged in the character of a reason, it forms a relevant and specific argument ; and belongs not to the account of fallacies ; and, if well founded, constitutes a just reason—if not for quashing the measure, at any rate for adding to the compensation thus shown to be deficient. And in this shape, viz. in that of a specific argument, will a man of course present his motive to view, if it be susceptible of it. But when the alleged damage and eventual injury will not, even in his own view of it, bear the test of inquiry, then, this specific argument failing him, he will betake himself to the general fallacy in lieu of it. He will set up the cry of *Innovation! Innovation!* hoping by this watchword to bring to his aid all whose sinister interest is connected with his own ; and to engage them to say, and the unreflecting multitude to believe, that the change in question is of the number of those in which the mischief attached to it is not accompanied by a preponderant mass of advantage.

Sect. 3. *Time the innovator-general, a counter-fal-*
lacy.

Among the stories current in the profession of the
law, is that of an attorney, who, when his client applied to him for relief against a forged bond, advised
him, as the shortest and surest course, to forge a release.

Thus, as a shorter and surer course than that of
attempting to make men sensible of the imposture,
this fallacy has been every now and then met by what
may be termed its counter-fallacy.—*Time itself is the*
arch-innovator. The inference is, the proposed change,
branded as it has thus been by the odious appellative
of innovation, is in fact no change; its sole effect
being either to prevent a change, or to bring the matter back to the good state in which it formerly was.
This counter-fallacy, if such it may be termed, has
not, however, any such pernicious properties or consequences attached to it as may be seen to be indicated by that name.   Two circumstances, however,
concur in giving it a just title to the appellation of a
fallacy : one is, that it has no specific application to
the particular measure in hand, and on that score
may be set down as irrelevant ; the other, that by a
sort of implied concession and virtual admission, it
gives colour and countenance to the fallacy to which
it is opposed : admitting by implication, that if the
appellation of a change belonged with propriety to
the proposed measure, it might on that single account
with propriety be opposed.

A few words, then, are now sufficient to strip the mask from this fallacy. No specific mischief, as likely to result from the specific measure, is alleged : if it were, the argument would not belong to this head. What is alleged is nothing more than that mischief, without regard to the amount, would be among the results of this measure. But this is no more than can be said of every legislative measure that ever did pass or ever can pass. If, then, it be to be ranked with arguments, it is an argument that involves in one common condemnation all political measures whatsoever, past, present, and to come ; it passes condemnation on whatsoever, in this way, ever has been or ever can be done, in all places as well as in all times. Delivered from an humble station, from the mouth of an old woman beguiling by her gossip the labours of the spinning-wheel in her cottage, it might pass for simple and ordinary ignorance :—delivered from any such exalted station as that of a legislative house or judicial bench, from such a quarter, if it can be regarded as sincere, it is a mark of *drivelling* rather than ignorance.

But it may be said, " My meaning is not to condemn all change, not to condemn all new institutions, all new laws, all new measures,—only violent and dangerous ones, such as that is which is now proposed." The answer is : Neither drawing or attempting to draw any line, you do by this indiscriminating appellative pass condemnation on all change ; on every thing to which any such epithet as *new* can with propriety be applied. Draw any such line, and the reproach of insincerity or imbecility shall be withholden : draw your line; but remember that whenever you do draw it, or so much as begin to draw it, you give up this your argument.

Alive to possible-imaginable evils, dead to actual ones,—eagle-eyed to future contingent evils, blind and insensible to all existing ones,—such is the character of the mind, to which a fallacy such as this can really have presented itself in the character of an argument possessing any the smallest claim to notice. To such a mind, that, by denial and sale of justice, anarchy, in so far as concerns nine-tenths of the people, is actually by force of law established, and that it is only by the force of morality, —of such morality as all the punishments denounced against sincerity, and all the reward applied for the encouragement of insincerity, have not been able to banish, that society is kept together ;—that to draw into question the fitness of great characters for their high situations, is in one man a crime, while to question their fitness so that their motives remain unquestioned is lawful to another ;—that the crime called *libel* remains undefined and undistinguishable, and the liberty of the press is defined to be the absence of that security which would be afforded to writers by the establishment of a licenser ;—that under a show of limitation, a government shall be in fact an absolute one, while pretended guardians are real accomplices, and at the nod of a king or a minister by a regular trained body of votes black shall be declared white ; miscarriage, success ; mortality, health ; disgrace, honour ; and notorious experienced imbecility, consummate skill ;—to such a mind, these, with other evils boundless in extent and number, are either not seen to be in existence, or not felt to be such.   In such a mind, the horror of innovation is as really a disease as any to which the body in which it is seated is exposed. And in proportion as a man is afflicted with it, he is the enemy of all good, which, how urgent soever may be the demand for it, remains as yet to be done; nor can he be said to be completely cured of it, till he shall have learnt to take on each occasion, and without repugnance, general utility for the general end, and, to judge of whatever is proposed, in the character of a means conducive to that end.

Sect. 4. *Sinister interests in which this fallacy has its source.*

Could the wand of that magician be borrowed at whose potent touch the emissaries of his wicked antagonist threw off their several disguises, and made instant confession of their real character and designs; —could a few of those ravens by whom the word *innovation* is uttered with a scream of horror, and the approach of the monster *Anarchy* denounced, —be touched with it, we should then learn their real character, and have the true import of these screams translated into intelligible language.

1. I am a lawyer (would one of them be heard to

say), a fee-fed judge, who, considering that the money
I lay up, the power I exercise, and the respect and
reputation I enjoy, depend on the undiminished continuance of the abuses of the law, the factitious delay,
vexation and expense with which the few who have
money enough to pay for a chance of justice are
loaded, and by which the many who have not, are cut
off from that chance,—take this method of deterring
men from attempting to alleviate those torments in
which my comforts have their source.

2. I am a sinecurist (cries another), who, being in
the receipt of 38,000*l.* a year, public money, for
doing nothing, and having no more wit than honesty,
have never been able to open my mouth and pronounce any articulate sound for any other purpose,
—yet, hearing a cry of " No sinecures !" am come to
join in the shout of " No innovation ! down with the
innovators !" in hopes of drowning, by these defensive
sounds, the offensive ones which chill my blood and
make me tremble.

3. I am a contractor (cries a third), who, having
bought my seat that I may sell my votes ; and in return for them, being in the habit of obtaining with
the most convenient regularity a succession of good
jobs, foresee, in the prevalence of innovation, the destruction and the ruin of this established branch of
trade.

4. I am a country gentleman (cries a fourth), who,
observing that from having a seat in a certain assembly a man enjoys more respect than he did before, on

the turf, in the dog-kennel, and in the stable, and having tenants and other dependents enough to seat me against their wills for a place in which I am detested, and hearing it said that if innovation were suffered to run on unopposed, elections would come in time to be as free in reality as they are in appearance and pretence,—have left for a day or two the cry of "Tally-ho!" and "Hark forward!" to join in the cry of "No Anarchy!" "No innovation!"

5. I am a priest (says a fifth), who, having proved the Pope to be Antichrist to the satisfaction of all Orthodox divines whose piety prays for the cure of souls, or whose health has need of exoneration from the burthen of residence; and having read, in my edition of the Gospel, that the apostles lived in palaces, which innovation and anarchy would cut down to parsonage-houses, though grown hoarse by screaming out, "No reading!" "No writing!" "No Lancaster!" and "No popery!"—for fear of coming change, am here to add what remains of my voice to the full chorus of "No Anarchy!" "No Innovation!"

CHAPTER III.

Fallacy of Distrust, or, What's at the bottom?
Ad metum.

Exposition.

This argument may be considered as a particular modification of the *No-Innovation* argument. An arrangement or set of arrangements has been proposed, so plainly beneficial, and at the same time so manifestly innoxious, that no prospect presents itself of bringing to bear upon them with any effect the cry of No innovation. Is the anti-innovationist mute? no: he has this resource :—In what you see as yet (says he) there may perhaps be no great mischief; but depend upon it, in the quarter from whence these proposed innoxious arrangements come, there are *more behind* that are of a very different complexion; if these innoxious ones are suffered to be carried, others of a noxious character will succeed without end, and will be carried likewise.

Exposure.

The absurdity of this argument is too glaring to be susceptible of any considerable illustration from any thing that can be said of it.

1. In the first place, it begins with a virtual admission of the propriety of the measure considered in itself; and thus, containing within itself a demonstra-

tion of its own futility, it cuts up from under it the very ground which it is endeavouring to make: yet, from its very weakness, it is apt to derive for the moment a certain degree of force.   By the monstrosity of its weakness, a feeling of surprise, and thereupon of perplexity, is apt to be produced : and so long as this feeling continues, a difficulty of finding an appropriate answer continues with it.   For that which is itself nothing, what answer (says a man) can I find?

2. If two measures, G and B, were both brought forward at the same time, G being good and B bad, rejecting G because B is bad would be quite absurd enough ; and at first view a man might be apt to suppose that the force of absurdity could go no further.

But the present fallacy does in effect go much further :—two measures, both of them brought upon the carpet together, both of them unobjectionable, are to be rejected, not for any thing that is amiss in either of them, but for something that by possibility may be found amiss in some other or others, that nobody knows of, and the future existence of which, without the slightest ground, is to be assumed and taken for granted.

In the field of policy as applied to measures, this vicarious reprobation forms a counterpart to vicarious punishment in the field of justice, as applied to persons.

The measure G, which is good, is to be thrown out, because, for aught we can be sure of, some day or other it may happen to be followed by some other measure B, which may be a bad one.   A man A, against whom there is neither evidence nor charge, is to be punished, because, for aught we can be sure of, some time or other there may be some other man who will have been guilty.

If on this ground it be right that the measure in question be rejected, so ought every other measure that ever has been or can be proposed : for of no measure can anybody be sure, but that it may be followed by some other measure or measures, of which, when they make their appearance, it may be said that they are bad.

If, then, the argument proves any thing, it proves that no measure ought ever to be carried, or ever to have been carried; and that, therefore, all things that can be done by law or government, and therefore law and government themselves, are nuisances.

This policy is exactly that which was attributed to Herod in the extermination of the innocents; and the sort of man by whom an argument of this sort can be employed, is the sort of man who would have acted as Herod did, had he been in Herod's place.

But think, not only what sort of man he must be who can bring himself to employ such an argument; but moreover, what sort of men they must be to whom he can venture to propose it; on whom he can expect it to make any impression, but such a one as will be disgraceful to himself. "Such drivellers," (says he to them in effect,) "such drivellers are you, so sure of being imposed upon, by any one that will attempt it, that you know not the distinction between good and bad : and when at the suggestion of this or that man you have adopted any one measure, good or bad, let but that same man propose any number of other measures, whatever be their character, ye are such idiots and fools, that without looking at them yourselves, or vouchsafing to learn their character from others, you will adopt them in a lump." Such is the compliment wrapt up in this sort of argument.

CHAPTER IV.

Official Malefactor's Screen.

Ad metum.

" Attack us, you attack Government."

Exposition.

THE fallacy here in question is employed almost
as often as, in speaking of the persons by whom, or
of the system on which, the business of the Government is conducted, any expressions importing condemnation or censure are uttered. The fallacy consists in affecting to consider such condemnation or
censure as being, if not in design, at least in tendency,
pregnant with mischief to government itself:—" Oppose us, you oppose Government;" " Disgrace us,
you disgrace Government;" " Bring us into contempt, you bring Government into contempt ; and
anarchy and civil war are the immediate consequences." Such are the forms it assumes.

Exposure.

Not ill-grounded, most assuredly, is the alleged
importance of this maxim : to the class of persons by
or for whom it is employed, it must be admitted to be
well worth whatsoever pains can be employed in deck-
ing it out to the best advantage.

Let but this notion be acceded to, all persons now
partaking, or who may at any time be likely to par-

take, in the business and profit of misrule, must, in every one of its shapes, be allowed to continue so to do without disturbance : all abuses, as well future as present, must continue without remedy. The most industrious labourers in the service of mankind will experience the treatment due to those to whose dissocial or selfish nature the happiness of man is an object of aversion or indifference. Punishment, or at least disgrace, will be the reward of the most exalted virtue ; perpetual honour, as well as power, the reward of the most pernicious vices. Punishment will be, and so by English libel-law it is at this day,—let but the criminal be of a certain rank in the state, and the mischief of the crime upon a scale to a certain degree extensive,—punishment will be, not for him who commits a crime, but for him who complains of it.

So long as the conduct of the business of the Government contains any thing amiss in it,—so long as it contains in it any thing that could be made better, —so long, in a word, as it continues short of a state of absolute perfection,—there will be no other mode of bringing it nearer to perfection, no other means of clearing it of the most mischievous abuses with which Government can be defiled, than the indication of such points of imperfection as at the time being exist, or are supposed to exist in it, which points of imperfection will always be referable to one or other of two heads :—the conduct of this or that one of the individuals by whom in such or such a department the business of Government is conducted ; or the state of

the system of administration under which they act.
But neither in the system in question, nor in the conduct of the persons in question, can any imperfection
be pointed out, but that, as towards such persons or
such system, in proportion to the apparent importance
and extent of that imperfection, aversion or contempt
must in a greater or less degree be produced.

In effect, this fallacy is but a mode of intimating
in other words, that no abuse ought to be reformed :
that nothing ought to be uttered in relation to the
misconduct of any person in office, which may produce
any sentiment of disapprobation.

In this country at least, few, if any persons, aim at
any such object as the bringing into contempt any of
those offices on the execution of which the maintenance
of the general security depends;—any such office,
for example, as that of king, member of parliament,
or judge. As to the person of the king, if the maxim,
"The king can do no wrong," be admitted in both its
senses, there can be no need of imputing blame to
him, unless in the way of defence against the impru-
dence or the improbity of those who, by groundless
or exaggerated eulogiums on the personal character
of the individual monarch on the throne, seek to ex-
tend his power, and to screen from censure or scrutiny the misconduct of his agents.

But in the instance of any other office, to reprobate
every thing, the tendency of which is to expose the
officer to hatred or contempt, is to reprobate every
thing that can be said or done, either in the way of

complaint against past, or for the purpose of preventing future transgressions;—to reprobate every thing
the tendency of which is to expose the office to hatred
or contempt, is to reprobate every thing that can be
said or done towards pointing out the demand for reform, how needful soever, in the constitution of the
office.

If, in the constitution of the office in respect of
mode of appointment, mode of remuneration, &c.,
there be any thing that tends to give all persons placed
in it an interest acting in opposition to official duty,
or to give an increased facility to the effective pursuit
of any such sinister interest, every thing that tends
to bring to view such sinister interest, or such facility,
contributes, it may be said, to bring the office itself into
contempt.

That under the existing system of judicature, so far
as concerns its higher seats, the interest of the judge
is, throughout the whole field of his jurisdiction, in a
state of constant and diametrical opposition to the
line of his duty;—that it is his interest to maintain
undiminished, and as far as possible to increase, every
evil opposite to the ends of justice, viz. uncertainty,
delay, vexation and expense;—that the giving birth
to these evils has at all times been more or less an
object with every judge (the present ones excepted, of
whom we say nothing) that ever sat on a Westminster
hall bench, and that under the present constitution of
the office it were weakness to expect at the hands of
a judge any thing better ;—whilst, that of the above-

mentioned evils, the load which is actually endured by
the people of this country, is, as to a very small part
only, the natural and unavoidable lot of human nature;
—are propositions which have already in this work
been made plain to demonstration, and in the belief
of which the writer has been confirmed by the observations of nearly sixty years;—propositions of the truth
of which he is no more able to entertain a doubt than
he is of his own existence.

But in these sentiments, has he any such wish as
to see enfeebled and exposed to effectual resistance
the authority of judges? of any established judicatory?
of any one occupier of any such judicial seat? No:
the most strenuous defender of abuse in every shape
would not go further than he in wishes, and upon occasion in exertion, for its support.

For preventing, remedying, or checking transgression on the part of the members of Government, or
preventing their management of the business of Government from becoming completely arbitrary, the
nature of things affords no other means than such, the
tendency of which, as far as they go, is to lower either
these managing hands, or the system, or both, in the
affection and estimation of the people: which effect,
when produced in a high degree, may be termed
bringing them into hatred and contempt.

But so far is it from being true that a man's aversion or contempt for the hands by which the powers
of Government, or even for the system under which
they are exercised, is a proof of his aversion or con-

tempt towards Government itself, that, even in proportion to the strength of that aversion or contempt, it is a proof of the opposite affection. What in consequence of such contempt or aversion he wishes for, is, not that there be no hands at all to exercise these powers, but that the hands may be better regulated; —not that those powers should not be exercised at all, but that they should be better exercised;—not that, in the exercise of them, no rules at all should be pursued, but that the rules by which they are exercised should be a better set of rules.

All government is a trust; every branch of government is a trust; and immemorially acknowledged so to be : it is only by the magnitude of the scale that public differ from private trusts.

I complain of the conduct of a person in the character of guardian, as domestic guardian, having the care of a minor or insane person. In so doing, do I say that guardianship is a bad institution? Does it enter into the head of any one to suspect me of so doing?

I complain of an individual in the character of a commercial agent, or assignee of the effects of an insolvent. In so doing, do I say that commercial agency is a bad thing? that the practice of vesting in the hands of trustees or assignees the effects of an insolvent for the purpose of their being divided among his creditors, is a bad practice? Does any such conceit ever enter into the head of man, as that of suspecting me of so doing?

I complain of an imperfection in the state of the law relative to guardianship. In stating this supposed imperfection in the state of the law itself, do I say that there ought to be no law on the subject? that no human being ought to have any such power as that of guardian over the person of any other? Does it ever enter into the head of any human being to suspect me so much as of entertaining any such persuasion, not to speak of endeavouring to cause others to entertain it?

Nothing can be more groundless than to suppose that the disposition to pay obedience to the laws by which security in respect of person, property, reputation and condition in life is afforded, is influenced by any such consideration as that of the fitness of the several functionaries for their respective trusts, or even so much as by the fitness of the system of regulations and customs under which they act.

The chief occasions in which obedience on the part of a member of the community in his character of subject is called upon to manifest itself, are the habitual payment of taxes, and submission to the orders of courts of justice : the one an habitual practice, the other an occasional and eventual one. But in neither instance in the disposition to obedience, is any variation produced by any increase or diminution in the good or ill opinion entertained in relation to the official persons by whom the business of those departments is respectively carried on, or even in relation to the goodness of the systems under which they act.

Were the business of Government carried on ever so much worse than it is, still it is from the power of Government in its several branches that each man receives whatsoever protection he enjoys, either against foreign or domestic adversaries. It is therefore by his regard for his own security, and not by his respect either for the persons by whom or the system according to which those powers are exercised, that his wish to see obedience paid to them by others, and his disposition to pay obedience to them himself, are produced.

Were it even his wish to withhold from them his own obedience, that wish cannot but be altogether ineffectual, unless and until he shall see others in sufficient number disposed and prepared to withhold each of them his own obedience; a state of things which can only arise from a common sense of overwhelming misery, and not from the mere utterance of complaint. There is no freedom of the press, no power to complain, in Turkey; yet of all countries it is that in which revolts and revolutions are the most frequent and the most violent.

Here and there a man of strong appetites, weak understanding, and stout heart excepted, it might be affirmed with confidence that the most indigent and most ignorant would not be foolish enough to wish to see a complete dissolution of the bonds of government. In such a state of things, whatsoever he might expect to grasp for the moment, he would have no assured hope of keeping. Were he ever so strong, his strength, he could not but see, would avail him nothing against a momentarily confederated multitude ; nor in one part of his field against a swifter individual ravaging the opposite part, nor during sleep against the weakest and most sluggish : and for the purpose of securing himself against such continually impending disasters, let him suppose himself entered into an association with others for mutual security; he would then suppose himself living again under a sort of government.

Even the comparatively few who, for a source of subsistence, prefer depredation to honest industry, are not less dependent for their wretched and ever palpitating existence than the honest and industrious are for theirs, on that general security to which their practice creates exceptions. Be the momentary object of his rapacity what it may, what no one of them could avoid having a more or less distinct conception of, is, that it could not exist for him further than it is secured against others.

So far is it from being true, that no Government can exist consistently with such exposure, no good Government can exist without it.

Unless by open and lawless violence, by no other means than lowering in the estimation of the people the hands by which the powers of Government are exercised, if the cause of the mischief consist in the unfitness of the hands ; or the system of management under which they act, if the cause of the mischief lie in the system,—be the hands ever so unfit, or the sy-

stem ever so ill-constructed,—can there be any hope
or chance of beneficial change.

There being no sufficient reason for ascribing even
to the worst-disposed any wish so foolish as that of
seeing the bonds of Government dissolved, nor on
the part of the best-disposed any possibility of contributing to produce change, either in any ruling hands
deemed by them unfit for their trust, or of the system
deemed by them ill adapted to those which are or
ought to be its ends, otherwise than by respectively
bringing into general disesteem these objects of their
disapprobation,—there cannot be a more unfounded
imputation or viler artifice, if it be artifice, or grosser
error, if it be error, than that which infers from the
disposition or even the endeavour to lessen in the estimation of the people the existing rulers, or the existing system, any such wish as that of seeing the
bands of Government dissolved.

In producing a local or temporary debility in the
action of the powers of the natural body, in many
cases, the honest and skilful physician beholds the
only means of cure : and from the act of the physician who precribes an evacuant or a sedative, it would
be as reasonable to infer a wish to see the patient
perish, as from the act of a statesman, whose endeavours are employed in lowering the reputation of the
official hands in whom, or the system of management
in which, he beholds the cause of what appears to
him amiss,—to infer a wish to see the whole frame of
Government either destroyed or rendered worse.

In so far as a man's feeling and conduct are influenced and determined by what is called *public opinion*, by the force of the *popular* or *moral sanction*,
and that opinion runs in conformity with the dictates
of the principles of general utility,—in proportion to
the value set upon reputation, and the degree of respect entertained for the community at large, his conduct will be the *better*, the more completely the quantity of respect he enjoys is dependent upon the goodness of his behaviour; it will be the *worse*, the more
completely the quantity of respect he is sure of enjoying is independent of it.

Thus, whatsoever portion of respect the people at
large are in the habit of bestowing upon the individual
by whom on any given occasion the office in question
is filled, this portion of respect may, so long as the
habit continues, be said to be attached to the office,
just as any portion of the emolument is which happens
to be attached to the office.

But as it is with emolument, so is it with respect.
The greater the quantity of it a man is likely to receive independently of his good behaviour, the less
good, in so far as depends upon the degree of influence
with which the love of reputation acts upon his mind,
is his behaviour likely to be.

If this be true, it is in so far the interest of the
public that that portion of respect, which along with
the salary is habitually attached to the office, should
be as small as possible.

If, indeed, the notion which it is the object of the

fallacy in question to inculcate were true, viz. that the stability of the Government or its existence at each given point of time depends upon the degree of respect bestowed upon the several individuals by whom at that point of time its powers are exercised,—if this were true, it would not be the interest of the public that the portion of respect habitually attached to the office, and received by the official person independently of his good behaviour in it, should be as small as possible. But in how great a degree this notion is erroneous has been shown already.

But while it is the interest of the public, that in the instance of each trustee of the public the remuneration received by him in the shape of respect should be as completely dependent as possible upon the goodness of his behaviour in the execution of his trust, it is the interest of the trustee himself that, as in every other shape, so in the shape of respect, whatsoever portion of the good things of this world he receives on whatever score, whether on the score of remuneration, or any other, should be as great as possible; since by good behaviour, neither respect nor any thing else can be always earned by him but by sacrifices in some shape or other, and in particular in the shape of ease.

Whatsoever, therefore, be the official situation which the official person in question occupies, it is his interest that the quantity of respect habitually attached to it be as great, and at the same time as securely attached to it, as possible.

And in the point of view from which he is by his personal and sinister interest led to consider the subject, the point of perfection in this line will not be attained until the quantity of respect he receives, in consequence of the possession he has of the office, be at all times as great as the nature of the office admits; —at all times as completely independent of the goodness of his behaviour in his office as possible ;—as great, in the event of his making the worst and least good use, as in that of his making the best and the least bad use, of the powers belonging to it.

Such being his interest, whatsoever be his official situation, if, as is the case of most if not all official situations, it be of such a nature as to have power in any shape attached to it, his endeavour and study will be so to order matters as to cause to be attached to it as above, and by all means possible, the greatest portion of respect possible.

To this purpose, amongst others, will be directed whatsoever influence his will can be made to act with on other wills, and whatsoever influence his understanding can be made to exert over other understandings.

If, for example, his situation be that of a judge; by the influence of will on will, it will seldom in any considerable degree be in his power to compell men by force to bestow upon him the sentiment of respect, either by itself or in any considerable degree by means of any external mark or token of it : but he may restrain men from saying or doing any of those things, the effect of which would be to cause others to bestow upon him less respect than they would otherwise.

If, being a judge of the King's Bench, any man has the presumption to question his fitness for such his high situation, he may for so doing punish him by fine and imprisonment with *et cæteras*.   If a Lord Chancellor, he may prosecute him before a judge, by whom a disposition to attach such punishments to such offences has been demonstrated by practice.

Thus much as to what can and what cannot be done towards attaching respect to office, by the influence of will on will.

What may be done by the influence of understanding on understanding remains to be noticed :—laying out of the question that influence which, in the official situation in question, is exercised over the understandings of the people at large independently of any exertions on the part of him by whom it is filled,—that which on his part requires exertion, and is capable of being exercised by exertion, consists in the giving utterance and circulation in the most impressive manner to the fallacy in question, together with a few such others as are more particularly connected with it.

Upon the boldness and readiness with which the hands and system are spoken ill of, depends the difference between arbitrary and limited government,—between a government in which the great body of the people have, and one in which they have not, a share.

In respect of the members of the governing body, undoubtedly the state of things most to be desired, is,

that the only occasion on which any endeavours should
be employed to lower them in the estimation of the
public should be those in which inaptitude in some
shape or other, want of probity, or weakness of judgment, or want of appropriate talent, have justly been
imputable to them : that on those occasions in which
inaptitude has not in any of those shapes been justly
imputable, no such endeavour should ever be employed.

Unfortunately, the state of things hereby supposed
is plainly (need it be said?) an impossible one. Admit no accusation, you may and you will exclude all
unjust ones;—admit just ones, you must admit unjust
ones along with them; there is no help for it. One
of two evils being necessarily to be chosen, the question is, which is the least?—to admit all such imputations, and thereby to admit of unjust ones, or to exclude all such imputations, and thereby to exclude all
just ones. I answer without difficulty,—the admission of unjust imputations is, beyond comparison, the
least of the two evils. Exclude all unjust imputations,
and with them all just ones, the only check by which
the career of deterioration can be stopped being thus
removed, both *hands* and *system* will, until they arrive
at the extreme of despotism and misrule, be continually
growing worse and worse;—the hands themselves will
grow worse and worse, having nothing to counteract
the force of that separate and sinister interest to the
action of which they remain constantly exposed;—
and the system itself will grow worse and worse, it

being all along the interest and, by the supposition,
within the power of the hands themselves to make
it so.

Admit just imputations, though along with them
you admit unjust ones, so slight is the evil as scarcely
to bear that name. Along with unjust imputations,
are not defences admitted? In respect of motives
and of means, have not the defendants in this case,
beyond all comparison, the advantage of the com-
plainants?

As far as concerns *motives*, in the instance of every
person included in the attack (and in an attack made
upon any one member of the Government as such,
who does not know how apt all are to feel themselves
included?), the principle of self-preservation is stronger
than the exciting cause productive of the disposition
to attack can be in any instance.

As far as concerns *means* of defence, if the person
against whom the attack is principally levelled wants
time or talent to defend himself, scarce a particle of
the immense mass of the matter of reward, which, in
all manner of shapes, for the purpose of carrying on
the ordinary business of government, lies constantly
at the disposal of the members of the Government,
but is applicable, even without any separate expense,
to the extraordinary purpose of engaging defending
advocates.

Let it not be said, " This is a persecution to which
an honourable man ought not to be exposed ;—a persecution which, though to some honourable men it

may be tolerable, will to others be intolerable,—intolerable to such a degree as to deprive the public of the benefit of their services."

A notion to any such effect will scarcely be advanced with a grave face.—That censure is the tax imposed by nature upon eminence, is the A B C of common place. Who is there to whom it can be a doubt that exposure to such imputations is among the inevitable appendages of office? If it were an office which in no shape whatever had any adequate allowance of the matter of reward annexed to it,—if it were a situation into which men were pressed,—the observation would have some better ground; but in the class of office here in question, exists there any such?

A self-contradiction is involved in the observation itself. The subject, of which sensibility thus morbid is predicated, is *an honourable man :* but to an honourable man, to any man to whom the attribute honourable can with truth and justice be applied, such sensibility cannot be attributed. The man who will not accept an office but upon condition that his conduct in it shall remain exempt from all imputation, intends not that his conduct shall be what it ought to be. The man to whom the idea of being subject to those imputations to which he sees the best are exposed, is intolerable,—is in his heart a tyrant,—and, to become so in practice, wants nothing but to be seated on one of those thrones, or on one of those benches, in which, by the appearance of chains made for show and not for use, a man is enabled, with the greater dignity as

well as safety, to act the part of the tyrant, and glut
himself with vengeance.

To a man who, in the civil line of office, accepts a
commission, it is not less evident that by so doing he
exposes himself to imputations, some of which may
happen to be unjust, than to a man in the military
line it is evident that by acceptance of a commission
in that line he exposes himself to be shot at: and of
a military office, with about equal truth, might it be
said, that an honourable man will not accept it on
such condition, as of a civil office that an honourable
man will not accept it, if his conduct is to stand exposed to such imputations.

In such circumstances, it is not easy to see how it
should happen to a public man to labour at the long-
run under an imputation that is not just. In so far as
any such incident does take place, evil does in truth
take place : but even in this case, the evil will not be
unaccompanied with concomitant good, operating in
compensation for it.   On the part of men in office, it
contributes to keep up the habit of considering their
conduct as exposed to scrutiny,—to keep up in their
minds that sense of responsibility on which goodness
of conduct depends, in which good behaviour finds its
chief security.

On the part of the people at large, it serves to keep
alive the expectation of witnessing such attacks ; the
habit of looking out for them ; and, when any such
attack does come, it prevents the idea of hardship
which is apt to attach upon any infliction, how neces-

sary soever, of which it can be said that it is unprecedented or even rare ; and hinders the public mind
from being set against the attack, and him who finds
exertion and courage enough to make it.

When, in support of such imputations, false facts
are alleged, the act of him by whom such false allegations are made, not only ought to be regarded as
pernicious, but ought to be, and is, consistently with
justice and utility, punishable;—punishable even when
advanced through temerity without consciousness of
the falsity, and more so when accompanied with such
dishonest consciousness.

But by a sort of law, of which the protection of
high-seated official delinquency is at least the effect,
not to say the object, a distinction thus obvious as
well as important has been carefully overlooked : and
whenever to the prejudice of the reputation of a man,
especially if he be a man in office, a fact which has
with more or less confidence been asserted or insinuated turns out to be false, the existence of dishonest
consciousness, whether really existing or not, is assumed.

In so far as public men, trustees and agents for the
people in possession or expectancy are the objects, a
general propensity to scrutinize into their conduct, and
thereby to cast imputations on it at the hazard of their
being more or less unmerited, is a useful propensity.
It is conducive to good behaviour on their part, and
for the opposite and corresponding reason, the habit of
general laudation, laudation without specific grounds,

is a mischievous propensity, being conducive to ill
behaviour on their part.

Render all such endeavours hopeless, you take from
a bad state of things all chance of being better :—allow to all such endeavours the freest range, you do no
injury to the best state of things imaginable.

Whatsoever facilities the adversaries of the existing
state of things have for lowering it in the estimation
of the people, equal facilities at least, if not greater,
have its friends and supporters for keeping and raising
it up.

Under the English constitution, at any rate, the
most strenuous defenders of the existing set of managing hands, as well as of the existing system of management, are not backward in representing an opposition as being no less necessary a power among the
springs of Government than the regulator in a watch[a].
But in what way is it that opposition, be it what it
may, ever acts or ever can act but by endeavouring
to lower either the managing hands, or, in this or that
part of it, the system of management, in the estimation
of the people ? and from a watchmaker's putting a
regulating spring into the watch he is making, it would
be just as reasonable and fair to infer that his meaning
is to destroy the watch, as from the circumstance of a
man's seeking, in this or that instance, to lower in the
estimation of the people the managing hands, or this
or that part of the system of management, to infer a
desire on his part to destroy the Government.

---

[a] More's *Observations*, p. 77, 78.

N

Under the English constitution at least, not only in point of fact is the disposition to pay that obedience by which the power of Government is constituted, and on which the existence of it depends, independent of all esteem for the hands by which this power is exercised, unaffected by any disesteem for this or that part of the system of management according to which it is executed,—but under such a constitution at least, the more complete this independence, the better for the stability and prosperity of the state. Being as it is, it suffices for carrying on at all times the business of Government,—viz. upon that footing in point of skill and prosperity which is consistent with the aptitude, probity and intelligence of the managing hands, and the goodness of the system of management under which they act : but if on each occasion it depended on the degree of estimation in which the conduct and character of the managing hands and the structure of the system of management under which they act happened at that time to be held by the majority of the people, this power would be seen strong, and perhaps too strong, at one time,—weak to any degree of weakness,—insufficient to any degree of insufficiency, —at another.

Among the peculiar excellencies of the English constitution, one is, that the existence of the Government, and even the good conduct of it, depends in a less degree than under any other monarchy upon the personal qualifications of the chief ruler, and upon the place he occupies in the estimation of the people. Conceive the character of the chief ruler perfect to a certain degree of perfection, all checks upon his power would be a nuisance. On the other hand, under a constitution of government into which checks upon that power are admitted, the stronger and more efficient those checks, the worse the personal character of the chief ruler may be, and the business of government still go on without any fatal disturbance.

On recent occasions, as if the endeavour had been new and altogether anomalous to the constitution, great were the outcries against the audacity of those parliamentary electors and other members of the community who, in the character of petitioners, were using their endeavours to lower the House of Commons in the estimation of the people, or, in stronger terms, to bring it and its authority into contempt. That by the individuals in question an endeavour of this nature should be regarded as a cause of personal inconvenience, and as such be resisted, is natural enough; but as to its being, on the part of the authors of those exertions, blameable,—or, on the part of the constitution, dangerous,—surely no further observation need here be added.

But what was complained of as an abuse, was the existence of that state of things, of that system of management, under which, in a number sufficient on ordinary occasions to constitute or secure a majority, the members of that governing body have a sinister interest separate from and opposite to that of the people for whom they profess to serve: that being in-

dependent as towards those to whom they ought to
be dependent,—as to those whom it is their duty to
control, and towards whom they ought to be independent,—they are dependent; and that, by means, by
which, though altogether out of the reach of punishment, the dependence is rendered beyond comparison
more constant and effectual than it would be by acts
of punishable bribery.

In this state of things, if any alteration in it be desirable, it is impossible that such alteration should be
brought about by other means than lowering in the
estimation of the people not only the system itself,
but all those who act willingly under it, and use their
endeavours to uphold it.

Without this means, and by any other means, how
is it that by possibility any such change should be
produced? Supposing them assured of possessing, in
the event of a refusal of all such change, as high a
place in the estimation of the people as they hold at
present, any thing done by them in furtherance of such
a change would be an effect without a cause. In their
personal capacities, they have all, or most of them,
little to gain, while they have much to lose, by any
proposed change.

True, it may be said, to be remedied, an imperfection, be it what it may, must be pointed out. But
what we complain of as dangerous to Government, is,
not the indication of such imperfections with their supposed remedies, but the mode in which they are apt
to be pointed out;—the heat, the violence, with which

uch indication is accompanied. This we object to,
not merely as dishonest, but as unwise,—as tending
to irritate the very persons at whose hands the remedy
thus pleaded for is sought.

To this, the answer is as follows :—

1. Whatsoever may be the terms most decorous,
and, upon the supposition, the best adapted to the
obtaining of the relief desired, it is not possible to
comprise them in any such scheme of description as
will enable a man to satisfy himself before-hand what
terms will be considered exposed to, what exempt
from, censure.

2. The cause of irritation is not so properly in the
terms of the application, as in the substance and nature
of the application itself : so that the greatest irritation
would be produced by that mode of application, whichever it were, that appeared most likely to produce the
effect in question;—the effect, the production of which
is on the one part an object of desire, on the other of
aversion : the least irritation by that which, in whatever terms couched, afforded the fairest pretence for
non-compliance.

3. The imperfection in question being, by the supposition, one of a public nature, the advantages of
which are enjoyed by a few, while the interest which
the many, each taken individually, have in the removal of the imperfection is commonly comparatively
small and remote, no little difficulty is commonly experienced by any one whose endeavour it should be
to persuade the many to collect amongst them a de-

gree of impressive force sufficient to operate upon the
ruling powers with effect.   On the part of the many,
the natural interest being in each case commonly but
weak, it requires to bring it into effective action whatsoever aids can be afforded it.   Strong arguments,
how strong soever, will of themselves be scarcely sufficient; for at the utmost they can amount to no more
than the indication of that interest which, in the case
of the greater part of the many whose force it is necessary to bring to bear upon the point in question, is
by the supposition but weak.   In aid of the utmost
strength of which the argument is susceptible, strength
of expression will therefore be necessary, or at least
naturally and generally regarded as necessary, and as
such employed.   But in proportion as this strength of
expression is employed, the mode of application stands
exposed to the imputation of that heat, and violence,
and acrimony, the use of which it is the object of the
alleged fallacy to prevent.

4. It is only on the supposition of its being in effect,
and being felt to be, conducive, or at least not repugnant, to the interest of the ruling powers addressed,
that the simple statement of the considerations which,
in the character of reasons, prove the existence of the
supposed imperfection, and, if a remedy be proposed,
the aptitude of the proposed remedy, can with reason
be expected to operate on them with effect.   But the
fact is, that on the part of those ruling powers, this
sort of repugnance, in a degree more or less considerable, is no other than what on every such occasion

ought in reason to be expected. If the imperfection in question be of the nature of those to which the term *abuse* is wont to be applied, these ruling powers have some or all of them, by the supposition, a special profit arising out of that abuse, a special interest consequently in the preservation and defence of it. Even if there be no such special interest, there exists in that quarter at all times, and in more shapes than one, a general and constant interest by which they are rendered mutually averse to applications of that nature. In the first place, in addition to their ordinary labours, they find themselves called upon to undertake a course of extraordinary labour, which it was not their design to undertake, and for which it may happen to some or all of them to feel themselves but indifferently prepared and qualified; and thus the application itself finds itself opposed by the interest of their ease. In the next place, to the extent of the task thus imposed upon them, they find the business of Government taken out of their hands. To that same extent their conduct is determined by a will, which originated not among themselves; and if, the measure being carried into effect, the promoters of it would obtain reputation, respect and affection, of those rewards a share more or less considerable falls into other hands : and thus the application in question finds an opponent in the interest of their pride.

CHAPTER V.

Accusation-scarer's Device.

Ad metum.

"Infamy must attach somewhere."

Exposition.

THIS fallacy consists in representing the imputation of purposed calumny as necessarily and justly attaching upon him who, having made a charge of misconduct against any person or persons possessed of political power or influence, fails of producing evidence sufficient for conviction.

Its manifest object, accordingly, is, as far as possible, to secure impunity to crimes and transgressions in every shape, on the part of persons so situated :—namely, by throwing impediments in the way of accusation, and in particular, by holding out to the eyes of those persons who have in view the undertaking the functions of accusers, in case of failure, in addition to disappointment, the prospect of disgrace.

Exposure.

"*Infamy must attach somewhere.*" To this effect was a *dictum* ascribed in the debates to the Right Honourable George Canning, on the occasion of the inquiry into the conduct of the Duke of York in his office of Commander in Chief.

In principle, insinuation to this effect has an un-limited application,—it applies, not only to all charges against persons possessed of political power, but, with more or less force, to all criminal charges in form of law against any persons whatsoever : and not only to all charges in a prosecution of the criminal cast, but to the litigants on both sides of the cause, in a case of a purely non-penal, or, as it is called, a civil nature.

If taken as a general proposition, applying to all public accusations, nothing can be more mischievous as well as fallacious. Supposing the charge unfounded, the delivery of it may have been accompanied with *mala fides* (consciousness of its injustice), *temerity* only, or it may have been perfectly blameless. It is in the first case alone that infamy can with propriety attach upon him who brings it forward. A charge really groundless may have been honestly *believed* to be well-founded, i. e. believed with a sort of pro-visional credence, sufficient for the purpose of en-gaging a man to do his part towards the bringing about an investigation, but without sufficient reasons. But a charge may be perfectly groundless without at-taching the smallest particle of blame upon him who brings it forward. Suppose him to have heard from one or more, presenting themselves to him in the cha-racter of percipient witnesses, a story, which, either *in toto*, or perhaps only in *circumstances*, though in circumstances of the most material importance, should prove false and mendacious,—how is the person who hears this, and acts accordingly, to blame? What sagacity can enable a man previously to legal investigation, a man who has no power that can enable him to ensure correctness or completeness on the part of this extra-judicial testimony, to guard against deception in such a case? Mrs. C. states to the accuser, that the Duke of York knew of the business; stating a conversation as having passed between him and herself on the occasion. All this (suppose) is perfectly false: but the falsity of it, how was it possible for one in the accuser's situation to be apprized of?

The tendency of this fallacy is, by intimidation, to prevent all true charges whatever from being made,—to secure impunity to delinquency in every shape.

But the conclusion, that because the discourse of a witness is false in one particular, or one occasion, it must therefore be false *in toto*,—in particular, that because it is false in respect of some fact or circumstance spoken to on some extra-judicial occasion, it is therefore not credible on the occasion of a judicial examination,—is a conclusion quite unwarranted.

If this argument were consistently and uniformly applied, no evidence at all ought ever to be received, or at least to be credited: for where was ever the human being, of full age, by whom the exact line of truth had never been in any instance departed from in the whole course of his life?

The fallacy consists, not in the bringing to view, as

lessening the credit due to the testimony of the witness, this or that instance of falsehood, as indicated by inconsistency or counter-evidence, but in speaking of them as *conclusive*, and as warranting the turning a deaf ear to every thing else the witness has said, or, if suffered, might have said. Under the pressure of some strong and manifest falsehood-exciting interest, suppose falsehood has been uttered by the witness : be it so ; does it follow that falsehood will on every occasion—will in the particular occasion in question— be uttered by him without any such excitement ?

Under the pressure of terror, the Apostle Peter, when questioned whether he were one of the adherents of Jesus, who at that time was in the situation of a prisoner just arrested on a capital charge,—denied his being so ; and in so doing, uttered a wilful falsehood; and this falsehood thrice repeated within a short time: —does it follow that the testimony of the Apostle ought not on any occasion to have been considered as capable of being true?

If any such rule were consistently pursued, what judge, who had ever acted in the profession of an advocate, could with propriety be received in the character of a witness ?

Again, with respect to the object of the charge, so far from receiving less countenance where the object is a public than where he is a private man, accusation, whether it be at the bar of an official judicatory or at the bar of the public at large, ought to receive, beyond comparison, more countenance. In case of the truth

of the accusation, the mischief is greater, the demand
for appropriate censure, as a check to it, correspondently greater. On the other hand, in case of non-de-
linquency, the mischief to the groundlessly-accused
individual is less. Power, in whatever hands lodged,
is almost sure to be more or less abused ; the check,
in all its shapes, so as it does not defeat the good
purposes for which the power has been given or suffered to be exercised, can never be too strong. That
against a man who, by the supposition, has done nothing wrong, it is not desirable, whether his situation
be public or private, that accusation should have been
preferred,—that he should have been subjected to the
danger, and alarm, and evil in other shapes attached
to it, is almost too plainly true to be worth saying.
But in the case of a public accusation, though, by the
supposition, it turns out to be groundless, it is not
altogether without its use ;— the evil produced is not
altogether without compensation : for by the alarm
it keeps up, in the breasts in which a disposition to
delinquency has place, such accusation acts as a
check upon it, and contributes to the prevention or
repression of it. On the other hand, in the situation
of the public man, the mischief, in the case of his
having been the object of an unfounded accusation, is
less, as we have shown in the preceding chapter,
than in the case of a private man. In the advantages that are attached to his situation, he possesses
a fund of compensation, which, by the supposition,
has no place in the other case : and apprized as he

ought to be, and, but for his own fault, is, of the enmity and envy to which, according to the nature of it, his situation exposes him, and not the private man, he ought to be, and, but for his own fault will be, proportionably prepared to expect it, and less sensibly affected by it when it comes.

PART THE THIRD.

FALLACIES OF DELAY,

The subject-matter of which is Delay in various
shapes ; and the object, to postpone discussion, with
a view of eluding it.

CHAPTER I.

The Quietist, or, " No Complaint."

Ad quietem.

Exposition.

A NEW law or measure being proposed in the
character of a remedy for some incontestable abuse
or evil, an objection is frequently started to the following effect :—" The measure is unnecessary ; nobody complains of disorder in that shape, in which it
is the aim of your measure to propose a remedy to it ;
even when *no* cause of complaint has been found to
exist, expecially under Governments which admit of
complaints, men have in general not been slow to
complain ; much less where any just cause of complaint has existed." The argument amounts to this :
—Nobody complains, therefore nobody suffers. It
amounts to a *veto* on all measures of precaution or

prevention, and goes to establish a maxim in legislation, directly opposed to the most ordinary prudence
of common life ;—it enjoins us to build no parapets
to a bridge till the number of accidents has raised an
universal clamour.

Exposure.

The argument would have more plausibility than it
has, if there were any chance of complaints being attended to ;—if the silence of those who suffer did not
arise from despair, occasioned by seeing the fruitless-
ness of former complaints.   The expense and vexation of collecting and addressing complaints to Parliament being great and certain, complaint will not
commonly be made without adequate expectation of
relief.   But how can any such expectation be entertained by any one who is in the slightest degree acquainted with the present constitution of Parliament?
Members who are independent of and irresponsible to
the people, can have very few and very slight motives
for attending to complaints, the redress of which
would affect their own sinister interests.   Again, how
many complaints are repressed by the fear of attacking powerful individuals, and incurring resentments
which may prove fatal to the complainant !

The most galling and the most oppressive of all
grievances is that complicated mass of evil which is
composed of the uncertainty, delay, expense and
vexation in the administration of justice : of this,
all but a comparatively minute proportion is clearly

factitious [a],—factitious, as being the work originally
and in its foundation of the man of law; latterly, and
in respect of a part of its superstructure, of the man
of finance. In extent, it is such, that of the whole
population, there exists not an individual who is not
every moment of his life exposed to suffer under it:
and few advanced in life, who, in some shape or other,
have not actually been sufferers from it. By the price
that has been put upon justice, or what goes by the
name of justice, a vast majority of the people, to some
such amount as $\frac{9}{10}$ths or $\frac{19}{20}$ths, are bereft altogether
of the ability of putting in for a chance for it; and to
those to whom, instead of being utterly denied this
sort of chance, it is sold, it is sold at such a price as
to the poorest of such as have it still in their power
to pay, the price is utter ruin, and even to the richest,
matter of serious and sensible inconvenience.

In comparison of this one scourge, all other political scourges put together are feathers: and in so far
as it has the operations of the man of finance for its
cause, if, instead of one-tenth upon income, a property
tax amounted to nine-tenths, still an addition to the
property tax would, in comparison of the affliction
produced by the sum assessed on law proceedings, be
a relief: for the income tax falls upon none but the
comparatively prosperous, and increases in proportion
to the prosperity, in proportion to the ability to sus-
tain it; whereas the tax upon law proceedings falls

---

[a] See Scotch Reform.

exclusively upon those whom it finds labouring under affliction,—under that sort of affliction which, so long as it lasts, operates as a perpetual blister on the mind.

Here, then, is matter of complaint for every British subject that breathes :—here, injustice, oppression and distress are all extreme : complaint there is none ; why ?—because by unity of sinister interest, and consequent confederacy between lawyer and financier, relief is rendered hopeless.

CHAPTER II.

Fallacy of False-consolation.

Ad quietem.

Exposition.

A MEASURE, having for its object the removal of some abuse, i. e. of some practice the result of which is, on the part of the many, a mass of suffering more than equivalent to the harvest of enjoyment reaped from it by the few, being proposed,—this argument consists in pointing to the general condition of the people in this or that other country, under the notion, that in that other country, either in the particular respect in question or upon the whole, the condition of the people is not so felicitous as, notwithstanding the abuse, it is in the country in and for which the measure of reform is proposed.

" What is the matter with you ?" " What would you have ?" Look at the people there, and there : think how much better off *you* are than *they* are. Your prosperity and liberty are objects of envy to them ;— your institutions are the models which they endeavour to imitate.

Assuredly, it is not to the disposition to keep an eye of preference turned to the bright side of things, where no prospect of special good suggests the opposite course,—it is not to such a disposition or such a habit that by the word *fallacy* it is proposed to affix a mark of disapprobation.

When a particular suffering, produced as it appears by an assignable and assigned cause, has been pointed out as existing, a man, instead of attending to it himself, or inviting to it the attention of others, employs his exertions in the endeavour to engage other eyes to turn themselves to any other quarter in preference (he being of the number of those whose acknowledged duty it is to contribute their best endeavours to the affording to every affliction within their view whatsoever relief may be capable of being afforded to it without preponderant inconvenience),— then, and then only, is it that the endeavour becomes a just ground for censure, and the means thus employed present a title to be received upon the list of *fallacies*.

Exposure.

The pravity as well as fallaciousness of this argument can scarcely be exhibited in a stronger or truer light than by the appellation here employed to characterize it.

1. Like all other fallacies upon this list, it is nothing to the purpose.

2. In his own case, no individual in his senses would accept it.   Take any one of the orators by whom this argument is tendered, or of the sages on whom it passes for sterling : with an observation of the general wealth and prosperity of the country in his

mouth instead of a half-year's rent in his hand, let any one of his tenants propose to pay him thus in his own coin,—will he accept it?

3. In a court of justice, in an action for damages, to learned ingenuity, did ever any such device occur as that of pleading assets in the hand of a third person, or in the hands of the whole country, in bar to the demand? What the largest wholesale trade is to the smallest retail, such and more in point of magnitude is the relief commonly sought for at the hands of the legislator, to the relief commonly sought for at the hands of the judge.—What the largest wholesale trade is to the smallest retail trade, such in point of magnitude, yea and more, is the injustice endeavoured at by this argument when employed in the seat of legislative power, in comparison of the injustice that would be committed by deciding in conformity to it in a court of justice.

No country so wretched, so poor in every element of prosperity, in which matter for this argument might not be found.

Were the prosperity of the country never so much greater than at present,—take for the country any country whatsoever, and for present time any time whatsoever,—neither the injustice of the argument, nor the absurdity of it, would in any the smallest degree be diminished.

Seriously and pointedly in the character of a bar, to any measure of relief, no, nor to the most trivial improvement, can it ever be employed. Suppose a bill brought in for converting an impassable road any where into a passable one, would any man stand up to oppose it who could find nothing better to urge against it than the multitude and goodness of the roads we have already? No: when in the character of a serious bar to the measure in hand, be that measure what it may, an argument so palpably inapplicable is employed, it can only be for the purpose of creating a diversion;—of turning aside the minds of men from the subject really in hand to a picture which by its beauty, it is hoped, may engross the attention of the assembly, and make them forget for the moment for what purpose they came there.

CHAPTER III.

Procrastinator's Argument.

Ad socordiam.

"Wait a little, this is not the time."

Exposition.

To the instrument of deception here brought to view, the expressions that may be given are various to an indefinite degree; but in its nature and conception nothing can be more simple.

To this head belongs every form of words by which, speaking of a proposed measure of relief, an intimation is given, that the time, whatever it be, at which the proposal is made, is too early for the purpose; and given, without any proof being offered of the truth of such intimation; such as, for instance, the want of requisite information, or the convenience of some preparatory measure.

Exposure.

This is the sort of argument or observation which we so often see employed by those who, being in wish and endeavour hostile to a measure, are afraid or ashamed of being seen to be so. They pretend, perhaps, to approve of the measure; they only differ as to the proper time of bringing it forward; but it may be matter of question whether, in any one instance, this observation was applied to a measure by a man whose wish it was not, that it should remain excluded for ever.

It is in legislation the same sort of quirk which in judicial procedure is called a plea in abatement. It has the same object, being never employed but on the side of a dishonest defendant, whose hope it is to obtain ultimate impunity and triumph by overwhelming his injured adversary with despair, impoverishment and lassitude.

A serious refutation would be ill bestowed upon so frivolous a pretence. The objection exists in the will, not in the judgment, of the objector. " Is it lawful to do good on the sabbath day?" was the question put by Jesus to the official hypocrites. Which is the properest day to do good? Which is the properest day to remove a nuisance? Answer, The very first day that a man can be found to propose the removal of it: and whosoever opposes the removal of it on that day, will, if he dare, oppose the removal on every other.

The doubts and fears of the parliamentary procrastinator are the conscientious scruples of his prototype the Pharisee, and neither the answer nor the example of Jesus has succeeded in removing these scruples. To him, whatsoever is too soon to-day, be assured that to-morrow, if not too soon, it will be too late.

True it is, that, the measure being a measure of reform or improvement, an observation to this effect may be brought forward by a friend to the measure: and in this case, it is not an instrument of deception, but an expedient of unhappily necessary prudence.

Whatsoever it may be some centuries hence, hitherto the fault of the people has been, not groundless clamour against imaginary grievances, but insensibility to real ones;—insensibility, not to the effect, the evil itself, for that, if it were possible, far from being a fault, would be a happiness,—but to the cause, to the system or course of misrule which is the cause of it.

What, therefore, may but too easily be—what hitherto ever has been—the fact, and that, throughout a vast proportion of the field of legislation, is, that in regard to the grievances complained of, the time for bringing forward a measure of effectual relief is not yet come: why? because, though groaning under the effect, the people, by the artifice and hypocrisy of their oppressors, having been prevented from entertaining any tolerably adequate conception of the cause, would at that time regard either with indifference or with suspicion the healing hand that should come forward with the only true and effectual remedy. Thus it is, for example, with that Pandora's box of grievances and misery, the contents of which are composed of the evils opposite to the ends of justice.

CHAPTER IV.

Snail's-pace Argument.

Ad socordiam.

" One thing at a time ! Not too fast ! Slow and sure !"

Exposition.

THE proposed measure being a measure of reform, requiring that for the completion of the beneficial work in question a number of operations be performed, capable, all or some of them, of being carried on at the same time, or successively without intervals, or at short intervals, the instrument of deception here in question consists in holding up to view the idea of graduality or slowness, as characteristic of the course which wisdom would dictate on the occasion in question. For more effectual recommendation of this course, to the epithet *gradual* are commonly added some such eulogistic epithets as *moderate* and *temperate ;* whereby it is implied, that in proportion as the pace recommended by the word *gradual* is quickened, such increased pace will justly incur the censure expressed by the opposite epithets,—immoderate, violent, precipitate, extravagant, intemperate.

Exposure.

This is neither more nor less than a contrivance for making out of a mere word an excuse for leaving un-

done an indefinite multitude of things which, the
arguer is convinced, and cannot forbear acknowledg-
ing, ought to be done.

Suppose half a dozen abuses which equally and
with equal promptitude stand in need of reform; this
fallacy requires, that without any reason that can be
assigned, other than what is contained in the pronouncing or writing of the word *gradual*, all but one
or two of them shall remain untouched.

Or, what is better, suppose that, to the effectual
correction of some one of these abuses, six operations
require to be performed—six operations, all which
must be done ere the correction can be effected,—to
save the reform from the reproach of being violent
and intemperate, to secure to it the praise of graduality, moderation and temperance, you insist, that of
these half-a-dozen necessary operations, some one or
some two only shall be talked of, and proposed to be
done ;—one, by one bill to be introduced this session
if it be not too late (which you contrive it shall be) ;
another, the next session ; which time being come,
nothing more is to be said about the matter, and there
it ends.

For this abandonment, no one reason that will bear
looking at can be numbered up, in the instance of any
one of the five measures endeavoured to be laid upon
the shelf; for if it could, that would be the reason
assigned for the relinquishment, and not this unmeaning assemblage of three syllables.

A suit which, to do full justice to it, requires but

six weeks, or six days, or six minutes in one day, has it been made to last six years? That your caution and your wisdom may not be questioned, by a first experiment reduce the time to five years, then if that succeeds in another parliament, should another parliament be in a humour (which it is hoped it will not), reduce it to four years,—then again to three years;—and if it should be the lot of your grandchildren to see it reduced to two years, they may think themselves well off, and admire your prudence.

Justice,—to which in every eye but that of the plunderer and oppressor, rich and poor have an equal right,—do nine-tenths of the people stand excluded from all hope of, by the load of expense that has been heaped up. You propose to reduce this expense. The extent of the evil is admitted, and the nature of the remedy cannot admit of doubt: but by the magic of the three syllables *gra-du-al*, you will limit the remedy to the reduction of about one-tenth of the expense. Some time afterwards you may reduce another tenth, and go on so, that in about two centuries, justice may, perhaps, become generally accessible.

Importance of the business—extreme difficulty of the business—danger of innovation—need of caution and circumspection—impossibility of foreseeing all consequences—danger of precipitation—every thing should be gradual—one thing at a time—this is not the time—great occupation at present—wait for more leisure—people well satisfied—no petitions presented—no complaints heard—no such mischief has yet

taken place—stay till it has taken place;—such is the
prattle which the magpye in office who, understanding
nothing, understands that he must have something to
say on every subject, shouts out among his auditors
as a succedaneum to thought.

Transfer the scene to domestic life, and suppose a
man who, his fortune not enabling him without running into debt to keep one race-horse, has been for
some time in the habit of keeping six. To transfer
to this private theatre the wisdom and the benefit of
the gradual system, what you would have to recommend to your friend would be something of this sort:
—Spend the first year in considering which of your
six horses to give up; the next year, if you can satisfy
yourself which it shall be, give up some one of them:
by this sacrifice, the sincerity of your intention and
your reputation for economy will be established; which
done, you need think no more about the matter.

As all psychological ideas have their necessary root
in physical ones, one source of delusion in psychological arguments consists in giving an improper extension
to some metaphor which has been made choice of.

It would be a service done to the cause of truth, if
some advocate for the gradual system would let us
into the secret of the metaphor or physical image, if
any, which he has in view, and in the same language
give us the idea of some physical disaster as the result
of precipitation. A patient killed by rapid bleeding,
a chariot dashed in pieces by runaway steeds, a vessel
overset by carrying too much sail in a squall,—all

these images suppose a degree of precipitation which, if pursued by the proposers of a political measure, would be at once apparent, and the obvious and as-signable consequence of their course would afford un-answerable arguments against them.

All this while though by a friend to the measure, no such word as above will be employed in the cha-racter of argument; yet cases are not wanting in which the dilatory course recommended may be consented to or even proposed by him.

Suppose a dozen distinct abuses in the seat of le-gislative power, each abuse having a set of members interested in the support of it ; attack the whole body at once, all these parties join together to a certainty, and oppose you with their united force. Attack the abuses one by one, and it is possible that you may have but one of these parties, or at least less than all of them, to cope with at a time. Possible ? Yes :— but of probability, little can be said. To each branch of the public service belongs a class of public servants, each of which has its sinister interest, the source of the mass of abuses on which it feeds ; and in the person and power of the universal patron, the fountain of all honour and of all abuse, all those sinister interests are joined and embodied into one.

This is a branch of science in which no man is ever deficient ; this is what is understood,—understood to perfection by him to whom nothing else ever was or can be clear,—*Hoc discunt omnes,* unto *alpha et beta puelli.*

If there be a case in which such graduality as is here described can have been consented to, and with a reasonable prospect of advantage, it must have been a case in which, without such consent, the whole business would be hopeless.

Under the existing system, by which the door of the theatre of legislation is opened by opulence to members, in whose instance application of the faculty of thought to the business about which they are supposed to occupy themselves, would have been an effect without a cause, so gross is the ignorance, and in consequence, even where good intention is not altogether wanting, so extreme the timidity and apprehension, that on their part, without assurance of extreme slowness, no concurrence to a proposal for setting one foot before another, at even the slowest pace, would be obtained at all: their pace, the only pace at which they can be persuaded to move, is that which the traveller would take, whose lot it should be to be travelling in a pitch-dark night, over a road broken and slippery, edged with precipices on each side. Time is requisite for quieting timidity: why? because time is requisite for instructing ignorance.

Sect. 1. *Lawyers ; their interest in the employment of this fallacy.*

In proportion to the magnitude of their respective shares in the general fund of abuse, the various fraternities interested in the support of abuses have each of them their interest in turning to the best account this as well as every other article in the list of fallacies.

But it is the fraternity of lawyers, who (if they have not decidedly the most to gain by the dexterous management of this or of other fallacies) have, from the greatest quantity of practice, derived the greatest degree of dexterity in the management of it.

Judicature requiring reflection, and the greater the complication of the case, the greater the degree and length of reflection which the case requires : under favour of this association, they have succeeded in establishing a general impression of a sort of proportion in quantity as well as necessity of connexion between delay and attention to justice. Not that, in fact, a hundredth part of the established delay has had any origin in a regard for justice; but,—for want of sufficient insight into that state of things by which in persons so circumstanced in power and interest the general prevalence of any such regard has been rendered physically impossible,—in his endeavours to propagate the notion of a sort of general proportion between delay and regard for justice, the man of law has, unhappily, been but too successful. And it is, perhaps, to this error, in respect to matters of fact, that the snail's-pace fallacy is indebted, more than to any other cause, for its dupes. Be this as it may, sure it is, that in no track of reform has the rate of progress, which it is the object of this fallacy to secure, been adhered to with greater effect. By the Statute Book, if run over, (and little more than the titles would be necessary) in this view, a curious exemplification of the truth of this observation is afforded. An abuse so monstrous, that, on the part of the judicial hands by which it was manufactured, the slightest doubt of the mischievousness of it was absolutely impossible,—generation after generation groaning under this abuse,—and at length, when, by causes kept of course as much as possible out of sight, the support of the abuse has been deemed no longer practicable, comes at length a remedy. And what remedy? Never any thing better than a feeble palliative.

CHAPTER V.

Fallacy of artful Diversion.

Ad verecundiam.

Exposition and Exposure.

THE device here in question may be explained by the following direction or receipt for the manufacture and application of it :—

When any measure is proposed, which on any account whatsoever it suits your interest or your humour to oppose, at the same time that, in consideration of its undeniable utility, or on any other account, you regard it unadvisable to pass direct condemnation on it,—hold up to view some other measure, such as, whether it bear any relation or none to the measure on the carpet, will, in the conception of your hearers, present itself as superior in the order of importance. Your language, then, is—Why that? (meaning the measure already proposed.)—Why not this?—or this? mentioning some other, which it is your hope to render more acceptable, and by means of it to create a diversion, and turn aside from the obnoxious measure the affections and attention of those whom you have to deal with.

One case there is, in which the appellation of fallacy cannot with justice be applied to this argument; and that is, where the effectuation or pursuit of the measure first proposed would operate as a bar or an

P

obstacle to some other measure of a more beneficial character held up to view by the argument as competitor to it: and what, in the way of Exposure, will be said of the sort of expedient just described, will not apply to this case.

However, where the measure first proposed is of unquestionable utility, and you oppose it merely because it is adverse to your own sinister interest, you must not suggest any relevant measure of reform in lieu of it, except in a case in which, in the shape of argument, every mode of opposition is considered as hopeless : for unless for the purpose of forestalling the time and attention that would be necessary to the effectuation of the proposed beneficial measure, a measure altogether irrelevant and foreign to it is set up, a risk is incurred, that something, however inferior in degree, may be effected towards the diminution of the abuse or imperfection in question.

In the character of an irrelevant counter-measure, any measure or accidental business whatever may be made to serve, so long as it can be made to preoccupy a sufficient portion of the disposable time and attention of the public men on whose suffrages the effectuation or frustration of the measure depends.

But supposing the necessity for a relevant counter-measure to exist, and you have accordingly given introduction to it, the first thing then to be done is, to stave off the undesirable moment of its effectuation as long as possible.

According to established usage, you have given no-

tice of your intention to propose a measure on the
subject and to the effect in question. The intention
is of too great importance to be framed and carried
into act in the compass of the same year or session :
you accordingly announce your intention for next session. When the next session comes, the measure is
of too great importance to be brought on the carpet
at the commencement of the session ; at that period
it is not yet mature enough. If it be not advisable to
delay it any longer, you bring it forward just as the
session closes. Time is thus gained, and without any
decided loss in the shape of reputation : for what you
undertook, has to the letter been performed. When
the measure has been once brought in, you have to
take your choice, in the first place, between operations
for delay and operations for rejection. Operations
for delay exhibit a manifest title to preference: so
long as their effect can be made to last, they accomplish their object, and no sacrifice either of design or
of reputation has been made. The extreme importance and extreme difficulty are themes on which you
blow the trumpet, and which you need not fear the
not hearing sufficiently echoed. When the treasury
of delay has been exhausted, you have your choice to
take between trusting to the chapter of accidents for
the defeat of the measure, or endeavouring to engage
some friend to oppose it and propose the rejection of
it. But you must be unfortunate indeed, if you can
find no opponents, no tolerably plausible opponents,
unless among friends, and friends specially commis-

sioned for the purpose: a sort of confidence more or less dangerous must in that case be reposed.

Upon the whole, you must however be singularly unfortunate or unskilful, if by the counter-measure of diversion any considerable reduction of the abuse or imperfection be, spite of your utmost endeavours, effected, or any share of reputation that you need care about, sacrificed.

PART THE FOURTH.

---◆---

FALLACIES OF CONFUSION,
The object of which is, to perplex, when Discussion can no longer be avoided.

CHAPTER I.

Question-begging Appellatives.

Ad judicium.

PETITIO *principii*, or begging the question, is a fallacy very well known even to those who are not conversant with the principles of logic. In answer to a given question, the party who employs the fallacy contents himself by simply affirming the point in debate. Why does opium occasion sleep?—Because it is soporiferous.

Begging the question is one of the fallacies enumerated by Aristotle; but Aristotle has not pointed out (what it will be the object of this chapter to expose) the mode of using the fallacy with the greatest effect, and least risk of detection,—namely, by the employment of a single appellative.

Exposition and Exposure.

Among the appellatives employed for the designation of objects belonging to the field of moral science, there are some by which the object is presented singly, unaccompanied by any sentiment of approbation or disapprobation attached to it :—as, *desire, labour, disposition, character, habit,* &c. With reference to the two sorts of appellatives which will come immediately to be mentioned, appellatives of this sort may be termed *neutral.*

There are others by means of which, in addition to the principal object, the idea of general approbation as habitually attached to that object is presented :— as, *industry, honour, piety, generosity, gratitude,* &c. These are termed *eulogistic* or laudatory.

Others there are again, by means of which, in addition to the principal object, the idea of general disapprobation, as habitually attached to that object, is presented :—as, *lust, avarice, luxury, covetousness, prodigality,* &c. These may be termed *dyslogistic* or vituperative [a].

Among pains, pleasures, desires, emotions, motives, affections, propensities, dispositions, and other moral entities, some, but very far from all, are furnished with appellatives of all three sorts :—some, with none but eulogistic ; others, and in a greater number, with none but those of the dyslogistic cast. By appella-

---

[a] See the nature of these denominations amply illustrated in Springs-of-Action Table.

Of the field of thought and action, this, the moral department, though it be that part in which the most abundant employment is given to the instrument of deception here in question, is not the only part. Scarcely, perhaps, can any part be found to which it has not been applied.

tives, I mean here, of course, *single-worded* appellatives; for by words, take but enough of them, any
thing may be expressed.

Originally, all terms expressive of any of these objects were (it seems reasonable to think) neutral. By
degrees they acquired, some of them an eulogistic,
some a dyslogistic, cast. This change extended itself
as the *moral sense* (if so loose and delusive a term
may on this occasion be employed) advanced in
growth.

But to return. As to the mode of employing this fallacy, it neither requires nor so much as admits of being
taught : a man falls into it but too naturally of himself; and the more naturally and freely, the less he
finds himself under the restraint of any such sense as
that of shame. The great difficulty is to unlearn it : in
the case of this, as of so many other fallacies, by teaching it, the humble endeavour here is to unteach it.

In speaking of the *conduct*, the *behaviour*, the *intention*, the *motive*, the *disposition* of this or that man,
if he be one who is indifferent to you, of whom you
care not whether he be well or ill thought of, you employ the *neutral* term :—if a man whom, on the occasion and for the purpose in question, it is your object
to recommend to favour, especially a man of your
own party, you employ the *eulogistic* term :—if he be
a man whom it is your object to consign to aversion
or contempt, you employ the *dyslogistic* term.

To the proposition of which it is the leading term,

every such eulogistic or dyslogistic appellative, secretly as it were, and in general insensibly, slips in another proposition of which that same leading term is the subject, and an assertion of approbation or disapprobation the predicate. The person, act, or thing in question is *or* deserves to be, or is *and* deserves to be, an object of general approbation; or the person, act, or thing in question is or deserves to be, or is and deserves to be, an object of general disapprobation.

The proposition thus asserted, is commonly a proposition that requires to be proved. But in the case where the use of the term thus employed is fallacious, the proposition is one that is not true, and cannot be proved : and where the person by whom the fallacy is employed is conscious of its deceptive tendency, the object in the employment thus given to the appellative is, by means of the artifice, to cause that to be taken for true which is not so.

By appropriate eulogistic and dyslogistic terms, so many arguments are made, by which, taking them altogether, misrule, in all its several departments, finds its justifying arguments, and these in but too many eyes, conclusive. Take, for instance, the following eulogistic terms :—

1. In the war department, *honour* and *glory*.

2. In international affairs, *honour, glory,* and *dignity*.

3. In the financial department, *liberality*. It being always at the expense of unwilling contributors that this *virtue* (for among the *virtues* it has its place in

Aristotle) is exercised; for *liberality, depredation* may, in perhaps every case, and without any impropriety, be substituted.

4. In the higher parts of all official departments, *dignity ;—dignity,* though not in itself depredation, operates as often as the word is used, as a pretence for, and thence as a cause of, depredation. Wherever you see *dignity,* be sure that money is requisite for the *support* of it : and that, in so far as the dignitary's own money is regarded as insufficient, public money raised by taxes imposed on all other individuals, on the principle of *liberality,* must be found for the supply of it [a].

Exercised at a man's own expense, liberality may be, or may not be, according to circumstances, a virtue :—exercised at the expense of the public, it never can be any thing better than vice. Exercised at a man's own expense, whether it be accompanied with prudence or no, whether it be accompanied or not with beneficence, it is at any rate disinterestedness :— exercised at the expense of the public, it is pure selfishness : it is, in a word, depredation : money or money's worth is taken from the public to purchase, for the use of the liberal man, respect, affection, gratitude with its eventual fruits in the shape of services of all sorts : in a word, reputation, power.

When you have a practice or measure to condemn,

---

[a] See this principle avowed and maintained by the scribes of both parties, Burke and Rose, as shown in the defences of economy against those advocates of depredation.

find out some more general appellative within the import of which the obnoxious practice or measure in question cannot be denied to be included, and to which you or those whose interests and prejudices you have espoused, have contrived to annex a certain degree of unpopularity, in so much that the name of it has contracted a dyslogistic quality,—has become a bad name.

Take, for example, *improvement* and *innovation ;* under its own name to pass censure on any improvement might be too bold : applied to such an object, any expressions of censure you could employ might lose their force : employing them, you would seem to be running on in the track of self-contradiction and nonsense.

But improvement means something new, and so does *innovation.* Happily for your purpose, *innovation* has contracted a bad sense ; it means something which is new and bad at the same time. Improvement, it is true, in indicating something new, indicates something good at the same time ; and therefore, if the thing in question be good as well as new, innovation is not a proper term for it. However, as the idea of *novelty* was the only idea originally attached to the term innovation, and the only one which is directly expressed in the etymology of it, you may still venture to employ the word innovation, since no man can readily and immediately convict your appellation of being an improper one upon the face of it.

With the appellation thus chosen for the purpose

of passing condemnation on the measure, he by whom
it has been brought to view in the character of an improvement is not (it is true) very likely to be well satisfied : but of this you could not have had any expectation.   What you want, is a pretence which your own
partisans can lay hold of, for the purpose of deducing
from it a colourable warrant for passing upon the improvement that censure which you are determined,
and they, if not determined, are disposed and intend,
to pass on it.

Of this instrument of deception, the potency is
most deplorable. It is but of late years that so much
as the nature of it has in any way been laid before
the public : and now that it has been laid before the
public, the need there is of its being opposed with
effect, and the extreme difficulty of opposing it with
effect, are at the same time and in equal degree manifest. In every part of the field of thought and discourse, the effect of language depends upon the principle of association,—upon the association formed between words and those ideas of which in that way
they have become the signs. But in no small part of
the field of discourse, one or other of the two censo-
rial and reciprocally correspondent and opposite affections,—the amicable and the hostile,—that by which
approbation and that by which disapprobation is expressed,—are associated with the word in question by
a tie little less strong than that by which the object
in question, be it person or thing,—be the thing a
real or fictitious entity, be it operation or quality, is

associated with that same articulate audible sign and its visible representations.

To diminish the effect of this instrument of deception (for to do it away completely, to render all minds without exception at all times insensible to it, seems scarcely possible), must, at any rate, be a work of time. But in proportion as its effect on the understanding, and through that channel on the temper and conduct of mankind, is diminished, the good effect of the exposure will become manifest.

By such of these passion-kindling appellatives as are of the eulogistic cast, comparatively speaking, no bad effect is produced : but by those which are of the dyslogistic, prodigious is the mischievous effect produced, considered in a moral point of view. By a single word or two of this complexion, what hostility has been produced ! how intense the feeling of it ! how wide the range of it ! how full of mischief, in all imaginable shapes, the effects [a] !

---

[a] As an instance remarkable enough, though not in respect of the mischievousness, yet in respect of the extent and the importance of the effects producible by a single word, note Lord Erskine's defence of the Whigs, avowedly produced by the application of the dyslogistic word *faction* to that party in the state.

CHAPTER II.

Impostor Terms.

Ad judicium.

Exposition.

THE fallacy which consists in the employment of impostor terms, in some respects resembles that which has been exposed in the preceding chapter : but it is applied chiefly to the defence of things, which under their proper name are manifestly indefensible: instead, therefore, of speaking of such things under their proper name, the sophist has recourse to some appellative, which, along with the indefensible object, includes some other ; generally an object of favour ;—or at once substitutes an object of approbation for an object of censure. For instance, persecutors in matters of religion have no such word as persecution in their vocabulary ; *zeal* is the word by which they characterize all their actions.

In the employment of this fallacy, two things are requisite :

1. A fact or circumstance, which, under its proper name, and seen in its true colours, would be an object of censure, and which, therefore, it is necessary to disguise ; (*res tegenda ;*)

2. The appellative, which the sophist employs to conceal what would be deemed offensive, or even to

bespeak a degree of favour for it by the aid of some happier accessary : (*Tegumen.*) [a]

Exposure.

Example.—*Influence of the Crown.*

The sinister influence of the crown is an object which, if expressed by any peculiar and distinctive appellation, would, comparatively speaking, find perhaps but few defenders, but which, so long as no other denomination is employed for the designation of it than the generic term *influence*, will rarely meet with indiscriminating reprobation.

---

[a] The device here in question is not peculiar to politicians. By an example drawn from private life, it may to some eyes be placed, perhaps, in a clearer point of view. The word *gallantry* is employed to denote either of two dispositions, which, though not altogether without connexion, may either of them exist without the other. In one of these senses, it denotes, on the part of the stronger sex, the disposition to testify on all occasions towards the weaker sex those sentiments of respect and kindness by which civilized is so strikingly and happily distinguished from savage life. In the other sense, it is, in the main, synonymous to *adultery*: yet, not so completely synonymous (as indeed words perfectly synonymous are of rare occurrence) but that, in addition to this sense, it presents an accessary and collateral one. Having, from the habit of being employed in the other sense, acquired, in addition to its direct sense, a collateral sense of the eulogistic cast, it serves to give to the act, habit, or disposition, which in this sense it is employed to present, something of an eulogistic tincture, in lieu of that dyslogistic colouring under which the object is presented by its direct and proper name. Whatever act a man regards himself as being known to have performed, or meditates the performance of, under any expectation of his being eventually known to have performed it, he will not, in speaking of it, make use of any term the tendency of which

Corruption,—the term which, in the eyes of those to whom this species of influence is an object of disapprobation, is the appropriate and only single-worded term capable of being employed for the expression of it,—is a term of the *dyslogistic* cast. This, then, by any person whose meaning it is not to join in the condemnation passed on the practice or state of things which is designated, is one that cannot possibly be employed. In speaking of this practice and state of things, he is therefore obliged to go upon the look-out, and find some term, which, at the same time that its claim to the capacity of presenting to view the object in question cannot be contested, shall be of the eulo-

---

is to call forth, on the part of the hearer or reader, any sentiment of disapprobation pointed at the sort of act in question, and consequently, through the medium of the act, at the agent by whom it has been performed. To the word *adultery*, this effect, to every man more or less unpleasant, is attached by the usage of language. On every occasion in which it is necessary to his purpose to bring to view an act of this obnoxious description, he will naturally be on the look-out for a term in the use of which he may be supposed to have had another meaning, and which, in so far as it conveys an idea of the forbidden act in question, presents it with an accompaniment, not of reproach, but rather of approbation, which in general would not have accompanied it but for the other signification which the word is also employed to designate. This term he finds in the word *gallantry*.

There is a sort of man, who, whether ready or no to commit any act or acts of adultery, would gladly be thought to have been habituated to the commission of such acts: but even this sort of man would neither be found to say of himself, " I am an adulterer," nor pleased to have it said of him, " He is an adulterer." But to have it said of him that he is a man of gallantry,—this is what the sort of man in question would regard as a compliment, with the sound of which he would be pleased and flattered.

gistic or at least of the neutral cast; and to one or
other of these classes belongs the term *influence.*

Under the term *influence,* when the crown is considered as the possessor of it, are included two species
of influence: the one of them, such, that the removal
of it could not, without an utter reprobation of the
monarchical form of Government, be by any person
considered as desirable, nor, without the utter destruction of monarchical government, be considered as possible. The other, such, that in the opinion of many
persons the complete destruction or removal of it
would if possible be desirable; and that, though consistently with the continuance of the monarchical government, the complete removal of it would not be
practicable, yet the diminution of it to such a degree
as that the remainder should not be productive of any
practically pernicious effects would not be impracticable.

Influence of *will on will,* influence of *understanding
on understanding :* in this may be seen the distinction
on which the utility or noxiousness of the sort of influence in question depends.

In the influence of understanding on understanding,
may be seen that influence to which, by whomsoever
exercised, on whomsoever exercised, and on what occasion soever exercised, the freest range ought to be
left :—left, although, as for instance, exercised by the
crown, and on the representatives of the people. Not
that to this influence it may not happen to be productive of mischief to any amount, but that because, with-

out this influence, scarce any good could be accomplished, and because, when it is left free, disorder cannot present itself without leaving the door open at least for the entrance of the remedy.

The influence of understanding on understanding is, in a word, no other than the influence of human reason,—a guide which, like other guides, is liable to miss its way, or dishonestly to recommend a wrong course, but which is the only guide of which the nature of the case is susceptible.

Under the British constitution, to the crown belongs either the sole management, or a principal and leading part of the management, of the public business: and it is only by the influence of understanding on understanding, or by the influence of will on will, that by any person or persons, except by physical force immediately applied, any thing can be done.

To the execution of the ordinary mass of duties belonging to the crown, the influence of will on will, so long as the persons on whom it is exercised are the proper persons, is necessary. On all persons to whom it belongs to the crown to give *orders*, this species of influence is necessary: for it is only in virtue of this species of influence that *orders*, considered as delivered from a superordinate to a subordinate,—considered in a word as *orders*, in contradistinction to mere suggestions, or arguments operating by the influence of understanding on understanding,—can be productive of any effect.

Thus far, then, in the case of influence of will on

Q

will, as well as in the case of influence of understanding on understanding, no rational and consistent objection can be made to the use of influence. In either
case, its title to the epithet *legitimate* influence is above
dispute.

The case among others in which the title of the influence of the crown is open to dispute,—the case in
which the epithet *sinister* or any other mark of disapprobation may be bestowed upon it (bestowed upon
the bare possession, and without need of reference to
the particular use and application which on any particular occasion may happen to be made of it),—is
that where, being of that sort which is exercised by
will on will, the person on whom on the occasion in
question it is exercised is either a member of parliament, or a person possessed of an electoral vote with
reference to a seat in parliament.

The ground on which this species of influence thus
exercised is, by those by whom it is spoken of with
disapprobation, represented as *sinister*, and deserving
of that disapprobation, is simply this :—viz. that in
so far as this influence is efficient, the will professed
to be pronounced is not in truth the will of him whose
will it professes to be, but the will of him in whom
the influence originates, and from whom it proceeds :
in so much, that if, for example, every member of
parliament without exception were in each house under
the dominion of the influence of the crown, and in
every individual instance that influence were effectual,
the monarchy, instead of being the limited sort of

monarchy it professes to be, would be in effect an absolute one,—in form alone a limited one; nor so much as in form a limited one any longer than it happened to be the pleasure of the monarch that it should continue to be.

The functions attached to the situation of a member of parliament may be included, most or all of them, under three denominations—the legislative, the judicial, and the inquisitorial : the legislative, in virtue of which, in each house, each member that pleases takes a part in the making of laws: the judicial, which, whether penal cases or cases non-penal be considered, is not exercised to any considerable extent but by the House of Lords : and the inquisitorial, the exercise of which is performed by an inquiry into facts, with a view to the exercise either of legislative authority, or of judicial authority, or both, whichever the case may be found to require. To the exercise of either branch may be referred what is done, when, on the ground of some defect either in point of moral or intellectual fitness, or both, application is made by either house for the removal of, any member or members of the executive branch of the official establishment, any servant or servants of the crown.

But for argument sake, suppose the above-mentioned extreme case to be realized, all these functions are equally nugatory. Whatever law is acceptable to the crown, will be not only introduced but carried. No law that is not acceptable to the crown, will be so much as introduced. Every judgment that is acceptable to the crown will be pronounced. No judgment that is not acceptable to the crown will be pronounced. Every inquiry that is acceptable to the crown will be made. No inquiry that is not acceptable to the crown will be made; and in particular, let, on the part of the servants of the crown, any or all of them, misconduct in every imaginable shape be ever so enormous, no application that is not acceptable to the crown will ever be made for their removal : that is, no such application will ever be made at all ; for in this state of things, supposing it in the instance of any servant of the crown to be the pleasure of the crown to remove him, he will be removed of course ; nor can any such application be productive of any thing better than needless loss of time.

Raised to the pitch supposed in this extreme case, there are not, it is supposed, many men in the country by whom the influence of the crown, of that sort which is exercised by the will of the crown on the wills of members of parliament, would not be really regarded as coming under the denomination of sinister influence ; not so much as a single one by whom its title to that denomination would be openly denied.

But among members of parliament, many there are on whom, beyond possibility of denial, this sort of influence—influence of will on will—is exerted : since no man can be in possession of any desirable situation from which he is removable, without its being exerted on him; say rather, without its exerting himself on him, for to the production of the full effect of influence, no act, no express intimation of will on the part of any person, is in any such situation necessary.

Here, then, comes the grand question in dispute. In some opinions, of that sort of influence of will on will, exercising itself from the crown on a member of parliament, or at any rate on a member of the House of Commons composed of the elected representatives of the people, not any the least particle is necessary, —not any the least particle is in any way beneficial, —not any the least particle, in so far as it is operative, can be other than pernicious.

In the language of those by whom this opinion is held, every particle of such influence is sinister influence, corrupt or corruptive influence, or in one word corruption.

Others there are in whose opinion, or at any rate, if not in their opinion, in whose language, of that influence thus actually exercising itself, the whole, or some part at any rate, is not only innoxious but beneficial, and not only beneficial but, to the maintenance of the constitution in a good and healthful state, absolutely necessary: and to this number must naturally be supposed to belong all those on whom this obnoxious species of influence is actually exercising itself.

CHAPTER III.

Vague Generalities.

Ad judicium.

Exposition.

VAGUE generalities comprehend a numerous class of fallacies resorted to by those who, in preference to the most particular and determinate terms and expressions which the nature of the case in question admits of, employ others more general and indeterminate.

An expression is vague and ambiguous when it designates, by one and the same appellative, an object which may be good or bad, according to circumstances; and if, in the course of an inquiry touching the qualities of such an object, such an expression is employed without a recognition of this distinction, the expression operates as a fallacy.

Take, for instance, the terms, *government, laws, morals, religion.* The *genus* comprehended in each of these terms may be divided into two species—the *good* and the *bad;* for no one can deny that there have been and still are in the world, bad governments, bad laws, bad systems of morals, and bad religions. The bare circumstance, therefore, of a man's attacking government or law, morals or religion, does not of itself afford the slightest presumption that he is engaged in any thing blameable : if his attack is only directed against that which is bad in each, his efforts may be productive of good to any extent.

This essential distinction the defender of abuse takes care to keep out of sight, and boldly imputes to his antagonist an intention to subvert all governments, laws, morals, or religion.

But it is in the way of insinuation, rather than in the form of direct assertion, that the argument is in this case most commonly brought to bear. Propose any thing with a view to the improvement of the existing practice in relation to government at large, to the law, or to religion, he will treat you with an oration on the utility and necessity of government, of law, or of religion. To what end? To the end that of your own accord you may draw the inference which it is his desire you should draw, even that what is proposed has in its tendency something which is prejudicial to one or other or all of these objects of general regard. Of the truth of the intimation thus conveyed, had it been made in the form of a direct assertion or averment, some proof might naturally have been looked for. By a direct assertion, a sort of notice is given to the hearer or reader to prepare himself for something in the shape of proof: but when nothing is asserted, nothing is on the one hand offered, nothing on the other expected, to be proved.

Exposure.

1. *Order.*

Among the several cloudy appellatives which have been commonly employed as cloaks for misgovernment, there is none more conspicuous in this atmosphere of illusion than the word *Order*.

The word *order* is in a peculiar degree adapted to the purpose of a cloak for *tyranny ;*—the word *order* is more extensive than law, or even than government.

But, what is still more material, the word *order* is of the eulogistic cast; whereas the words *government* and *law*, howsoever the things signified may have been taken in the lump for subjects of praise, the complexion of the signs themselves is still tolerably neutral : just as is the case with the words *constitution* and *institutions*.

Thus, whether the measure or arrangement be a mere transitory measure or a permanent law, if it be a tyrannical one, be it ever so tyrannical, in the word *order* you have a term not only wide enough, but in every respect better adapted than any other which the language can supply, to serve as a cloak for it. Suppose any number of men, by a speedy death or a lingering one, destroyed for meeting one another for the purpose of obtaining a remedy for the abuses by which they are suffering, what nobody can deny is, that by their destruction, *order* is maintained ; for the *worst* order is as truly *order* as the *best*. Accordingly, a clearance of this sort having been effected, suppose in

the House of Commons a Lord Castlereagh, or in
the House of Lords a Lord Sidmouth, to stand up
and insist that by a measure so undeniably prudential
order was maintained, with what truth could they be
contradicted ? And who is there that would have the
boldness to assert that order ought not to be maintained ?

To the word *order* add the word *good*, the strength
of the checks, if any there were, that were thus applied to tyranny would be but little if at all increased.
By the word *good*, no other idea is brought to view
than that of the sentiment of approbation, as attached
by the person by whom it is employed to the object
designated by the substantive to which this adjunct is
applied. Order is any arrangement which exists with
reference to the object in question ;—good order is
that order, be it what it may, which it is my wish to
be thought to approve of.

Take the state of things under *Nero*, under *Caligula :* with as indisputable propriety might the word
order be applied to it as to the state of things at
present in Great Britain or the American United
States.

What in the eyes of Bonaparte was good order ?—
That which it had been his pleasure to establish.

By the adjunct *social*, the subject *order* is perhaps
rendered somewhat the less fit for the use of tyrants,
but not much. Among the purposes to which the
word *social* is employed, is indeed that of bringing to
view a state of things favourable to the happiness of

society : but a purpose to which it is also employed, is that of bringing to view a state of things no otherwise considered than as having place in society. By the war which, in the Roman history, bears the name of the social war, no great addition to the happiness of society was ever supposed to be made, yet it was not the less a social one.

As often as any measure is brought forward having for its object the making any the slightest defalcation from the amount of the sacrifice made of the interest of the many to the interest of the few, *social* is the adjunct by which the *order* of things to which it is pronounced hostile is designated.

By a defalcation made from any part of the mass of factitious delay, vexation and expense, out of which and in proportion to which lawyers' profit is made to flow,—by any defalcation made from the mass of needless and worse than useless emolument to office, with or without service or pretence of service,—by any addition endeavoured to be made to the quantity or improvement in the quality of service rendered, or time bestowed in service rendered in return for such emolument,—by every endeavour that has for its object the persuading the people to place their fate at the disposal of any other agents than those in whose hands breach of trust is certain, due fulfilment of it morally and physically impossible,—*social order* is said to be endangered, and threatened to be destroyed.

Proportioned to the degree of clearness with which

the only true and justifiable end of government is held
up to view in any discourse that meets the public eye,
is the danger and inconvenience to which those rulers
are exposed, who, for their own particular interest,
have been engaged in an habitual departure from that
only legitimate and defensible course. _ Hence it is,
that, when compared with the words *order, maintenance of order*, the use even of such words as *happiness, welfare, well-being*, is not altogether free from
danger, wide-extending and comparatively indeterminate as the import of them is : to the single word *happiness* substitute the phrase *greatest happiness of the
greatest number*, the description of the end becomes
more determinate and even instructive, the danger
and inconvenience to misgovernment, and its authors,
and its instruments still more alarming and distressing; for then, for a rule whereby to measure the goodness or badness of a government, men are referred to
so simple and universally apprehensible a standard as
the numeration table.   By the pointing men's atten-
tions to this end, and the clearness of the light thus
cast upon it, the importance of such words as the
word *order*, which by their obscurity substitute to the
offensive light the useful and agreeable darkness, is
more and more intimately felt.

2. *Establishment.*

In the same way again, *Establishment* is a word in
use, to protect the bad parts of establishments, by

charging those who wish to remove or alter them with the wish to subvert all establishments, or all good establishments[a].

3. *Matchless Constitution.*

The constitution has some good points; it has some bad ones: it gives facility and, until reform—radical reform, shall have been accomplished, security and continual increase to waste, depredation, oppression and corruption in every department, and in every variety of shape.

Now in their own name respectively, waste, depredation, oppression, corruption, cannot be toasted: gentlemen would not cry, Waste for ever! Depredation for ever! Oppression for ever! Corruption for ever! But The constitution for ever! this a man may cry, and does cry, and makes a merit of it.

---

[a] In the church establishment, the bad parts are:—

1. Quantity and distribution of payment;—its inequality creating opposite faults—excess and deficiency. The excessive part calling men off from their duty, and, as in lotteries, tempting an excessive number of adventurers : the defect deterring men from engaging in the duty, or rendering them unable to perform it as it ought to be performed.

2. Mode of payment;—tithes, a tax on food, which discourages agricultural improvements, and occasions dissention between the minister and his parishioners.

3. Forms of admission, compelling insincerity, subversive of the basis of morality. As to purely speculative points, no matter which side a man embraces, so he be *sincere*, but highly mischievous that he should maintain even the right side (where there happens to be any) when he is *not* sincere.

Of this instrument of rhetoric, the use is at least as old as Aristotle. As old as Aristotle is even the receipt for making it; for Aristotle has himself given it: and of how much longer standing the use of it may have been, may baffle the sagacity of a Mitford to determine. How sweet are gall and honey? how white are soot and snow?

Matchless constitution! there's your sheet-anchor! there's your true standard! rally round the constitution: that is, rally round waste, rally round depredation, rally round oppression, rally round corruption, rally round election terrorism, rally round imposture; —imposture on the hustings, imposture in honourable house, imposture in every judicatory.

Connected with this toasting and this boasting is a theory, such as a Westminster or Eton boy on the sixth form, aye, or his grandmother, might be ashamed of. For among those who are loudest in crying out theory (as often as any attempt is made at reasoning, any appeal made to the universally known and indisputable principles of human nature), always may some silly sentimental theory be found.

The constitution, why must it not be looked into? why is it that under pain of being *ipso facto* anarchist convict, we must never presume to look at it otherwise than with shut eyes? Because it was the work of our ancestors :—of ancestors, of legislators, few of whom could so much as read, and those few had nothing before them that was worth the reading. First theoretical supposition, *wisdom of barbarian ancestors.*

When from their ordinary occupation, their order of the day, the cutting of one another's throats, or those of Welchmen, Scotchmen, or Irishmen, they could steal now and then a holiday, how did they employ it? In cutting Frenchmen's throats in order to get their money; this was active virtue :—leaving Frenchmen's throats uncut was indolence, slumber, inglorious ease. Second theoretical supposition, *virtue of barbarian ancestors.*

Thus fraught with habitual wisdom and habitual virtue, they sat down and devised; and setting before them the best ends, and pursuing those best ends by the best means, they framed in outline, at any rate, they planned and executed our matchless constitution: —the constitution as it stands,—and may it for ever stand!

Planned and executed? On what occasion? on none. At what place? at none. By whom? by nobody.

At no time? Oh yes, says every-thing-as-it-should-be Blackstone. Oh yes, says Whig after Whig, after the charming commentator, anno Domini 1660, then it is that it was in its perfection, about fourteen years before James the Second mounted the throne with a design to govern in politics as they do in Morocco, and in religion as they do at Rome; to govern without parliament, or in spite of parliament : a state of things for which, at this same era of perfection, a preparation was made by a parliament, which, being brought into as proper a state of corruption as if Lord

Castlereagh had had the management of it, was kept on foot for several years together, and would have been kept a-foot till the whole system of despotism had been settled, but for the sham popish plot by which the fortunate calumny and subornation of the Whigs defeated the bigotry and tyranny of the Tories.

What then says the only true theory,— that theory which is uniformly confirmed by all experience ?

On no occasion, in no place, at no time, by no person possessing any adequate power, has any such end in view as the establishing the greatest happiness of the greatest number, been hitherto entertained : on no occasion on the part of any such person has there been any endeavour, any wish for any happiness, other than his own and that of his own connexions, or any care about the happiness or security of the subject many, any further than his own has been regarded as involved in it.

Among men of all classes, from the beginning of those times of which we have any account in history, among all men of all classes, an universal struggle and contention on the part of each individual for his own security and the means and instruments of his own happiness, for money, for power, for reputation natural and factitious, for constant ease, and incidental vengeance.   In the course of this struggle, under favourable circumstances connected with geographical situation, this and that little security has been caught at, obtained, and retained by the subject many, against the conjoined tyranny of the monarch and his aristocracy.   No plan pursued by any body at any time;—the good established, as well as the bad remaining, the result of an universal scramble, carried on in the storm of contending passions under favour of opportunity:—at each period, some advantages which former periods had lost, others, which they had not gained.

But the only regular and constant means of security being the influence exercised by the will of the people on the body which in the same breath admit themselves and deny themselves to be their agents, and that influence having against it and above it the corruptive and counter-influence of the ruling few, the servants of the monarchy and the members of the aristocracy,—and the quantity of the corruptive matter by which that corruptive influence operates being every day on the increase, hence it is, that while all names remain unchanged, the whole state of things grows every day worse and worse, and so will continue to do till even the forms of parliament are regarded as a useless incumbrance, and pure despotism, unless arrested by radical reform, takes up the sceptre without disguise.

While the matter of waste and corruption is continually accumulating, while the *avalanche* composed of it is continually rolling on, that things should continue long in their present state seems absolutely impossible. Three states of things contend for the ultimate result:—Despotic monarchy undisguised by form; Representative democracy under the form of monarchy; Representative democracy under its own form.

In this, as in every country, the Government has been as favourable to the interests of the ruling few, and thence as unfavourable to the general interests of the subject many, or in one word as *bad*, as the sub ject many have endured to see it, have persuaded themselves to suffer it to be.   No abuse has, except under a sense of necessity, been parted with : no remedy, except under the like pressure, applied.   But under the influence of circumstances in a great degree peculiar to this country, at one time or another the ruling few have found themselves under the necessity of sacrificing this or that abuse, of instituting or suffering to grow up this or that remedy.

It is thus, that under favour of the contest between Whigs and Tories, the liberty of the press, the foundation of all other liberties, has been suffered to grow up and continue.   But this liberty of the press is not the work of institution, it is not the work of law; what there is of it that exists, exists not by means, but in spite, of law.   It is all of it contrary to law ; by law there is no more liberty of the press in England, than in Spain or Morocco.   It is not the constitution of the Government, it is not the force of the law; it is the weakness of the law we have to thank for it.   It is not the Whigs that we have to thank for it, any more than the Tories.   The Tories, that is, the supporters of monarchy, would destroy it, simply assured of their never being in a condition to have need of it.   The Whigs would with equal readiness destroy it, or concur in destroying it, could they possess that same comfortable assurance. But it has never been in their power: and to that impotence is it that we are indebted for their zeal for the liberty of the press and the support they have given to the people in the exercise of it. Without this arm they could not fight their battles: without this for a trumpet they could not call the people to their aid.

Such corruption was not, in the head of any original framer of the constitution, the work of design: but were this said without explanation, an opinion that would naturally be supposed to be implied in it, is, that the constitution was originally in some one head, the whole, or the chief part of it, the work of design. The evil consequence of a notion pronouncing it the work of design, would be, that, such a design being infinitely beyond the wisdom and virtue of any man in the present times, a planner would be looked out for in the most distant age that could be found,— thus the ancestor-wisdom fallacy would be the ruling principle, and the search would be fruitless and endless. But the non-existence of any determinate design in the formation of the constitution may be proved from *history*. The House of Commons is the characteristic and vital principle. Anno 1258 the man by whom the first *germ* was planted, was Simon de Montfort, Earl of Leicester, a foreigner and a rebel. In this first call to the people there was no better nor steadier design than that of obtaining momentary support for rebellion. The practice of seeing and hearing deputies from the lower orders before money was attempted to be taken out of their pockets, having thus sprung up, in the next reign Edward the First saw his convenience in conforming to it. From this time till Henry the Sixth's, instances in which laws were enacted by kings, sometimes without consulting commons, sometimes without consulting them or lords, are not worth looking out. Henry the Sixth's was the first reign in which the House of Commons had really a part in legislation ; till then they had no part in the penning of any laws ; no law was penned till after they were dissolved : here then, so late as about 1450 (between 1422 and 1459), the House of Commons as a branch of the legislature was an innovation, till then, (Anno 1450) *constitution* (if the House of Commons be a part of it) there was none. Parliament ? yes : consisting of king and lords, *legislators* ; deputies of commons, *petitioners*. Even of this aristocratical parliament the existence was precarious : indigence or weakness produced its occasional reproduction ; more prudence and good fortune would have sufficed for throwing it into disuse and oblivion : like the obsolete legislative bodies of France and Spain, it would have been reduced to a possibility. All this while, and down to the time when the re-assembling of parliaments was imperfectly secured by indeterminate laws occasioned by the temporary nature of pecuniary supplies, and the constant cra-

vings of royal paupers ; had the constitution been a
tree, and both houses branches, either or both might
have been lopped off, and the tree remain a tree still.[a]

After the bloody reigns of Henry the Eighth, and
Mary, and the too short reign of Edward the Sixth,
comes that of Elizabeth, who openly made a merit of
her wish to govern without parliament: members pre-
suming to think for themselves, and to speak as they
thought, were sent to prison for repentance. After
the short parliaments produced in the times of James
the First, and Charles the First, by profusion and distress, came the first long parliament. Where is now
the constitution ? Where the design ?—the wisdom ?
—The king having tried to govern without lords or
commons, failed : the commons having extorted from
the king's momentary despair, the act which converted
them into a perpetual aristocracy, tried to govern
without king or lords, and succeeded. In the time
of Charles the Second, no design but the king's design of arbitrary government executed by the instru-
mentality of seventeen years long parliament. As
yet, for the benefit of the people, no feasible design,

---

[a] Between Henry the Third, and Henry the Sixth (Anno 1258 to 1422)
it is true there were frequent acts ordaining annual and even oftener
than annual parliaments[*]. Still these were but vague promises, made
only by the king, with two or three petty princes: the commons were
not legislators, but petitioners : never seeing till after enactment the
acts to which their assent was recorded.

---

[*] See Christian on Blackstone.

but in the seat of supreme power; and *there*, conception of any such design scarce in human nature.

The circumstance to which the cry of matchless constitution is in a great degree indebted for its pernicious efficiency, is—that there was a time in which the assertion contained in it was incontrovertibly true. Till the American colonies threw off the yoke and became independent states; no political state possessed of a constitution, equalling it or approaching it in goodness, was any where to be found.

But from this its goodness in a comparative state, no well-grounded argument could at any time be afforded against any addition that could at any time be made to its intrinsic goodness. Persons happier than myself are not to be found any where: in this observation, supposing it true, what reason is there for my forbearing to make myself as much happier than I am at present, as I can make myself?

This pre-eminence is therefore nothing to the purpose; for of the pains taken in this way to hold it up to view, the design can be no other than to prevent it from being ever greater than it is.

But another misfortune is, that it is every day growing less and less: so that while men keep on vaunting this spurious substitute to positive goodness, sooner or later it will vanish altogether.

The supposition always is—that it is the same one day as another. But never for two days together has this been true. Since the revolution took place, never, for two days together, has it been the same:

every day it has been worse than the preceding : for
by every day, in some way or other, addition has
been made to the quantity of the matter of corruption—to that matter by which the effect of the only
efficient cause of good government, the influence of
the people, has been lessened.

A pure despotism may continue in the same state
from the beginning to the end of time : by the same
names the same things may be always signified. But
a mixt monarchy such as the English never can continue the same : the names may continue in use for
any length of time ; but by the same names the same
state of things is never, for two days together, signified. The quantity of the matter of corruption in the
hands of the monarch being every day greater and
greater ; the practice in the application of it to its
purpose, and thence the skill with which application
is made of it on the one hand, and the patience and
indifference with which the application of it is wit-
nessed, being every day greater and greater, the comparative quantity of the influence of the people, and
of the security it affords, is every day growing less and
less.

While the same names continue, no difference in
the things signified is ever perceived, but by the very
few who, having no interest in being themselves deceived, nor in deceiving others, turn their attention to
the means of political improvement. Hence it was,
that with a stupid indifference or acquiescence the
Roman people sat still, while their constitution, a bad

and confused mixture of aristocracy and democracy, was converted into a pure despotism.

With the title of representatives of the people, the people behold a set of men meeting in the House of Commons, originating the laws by which they are taxed, and concurring in all the other laws by which they are oppressed. Only in proportion as these their nominal representatives are chosen by the free suffrages of the people, and, in case of their betraying the people, are removable by them, can such representatives be of any use. But except in a small number of instances,—too small to be on any one occasion soever capable of producing any visible effect,—neither are these pretended representatives ever removable by them, nor have they ever been chosen by them. If, instead of a House of Commons and a House of Lords, there were two Houses of Lords, and no House of Commons, the ultimate effect would be just the same. If it depended on the vote of a reflecting man whether, instead of the present House of Commons, there should be another House of Lords, his vote would be for the affirmative: the existing delusion would be completely dissipated, and the real state of the nation be visible to all eyes; and a deal of time and trouble which is now expended in those debates, which, for the purpose of keeping on foot the delusion, are still suffered, would be saved.

As to representation, no man can even now be found so insensible to shame, as to affirm that any real representation has place: but though there is no real representation, there is, it is said, a *virtual* one : and with this, those who think it worth their while to keep up the delusion, and those who are, or act and speak as if they were, deluded, are satisfied. If those who are so well satisfied with a virtual representation which is not real, would be satisfied with a like virtual receipt of taxes on the one part, and a virtual payment of taxes on the other, all would be well ; but this unfortunately is not the case. The payment is but too real, while the falsity of the only ground on which the exaction of it is so much as pretended to be justified, is matter of such incontestable verity, and such universal notoriety, that the assertion of its existence is a cruel mockery.

4. *Balance of Power.*

In general, those by whom this phrase has been used, have not known what they meant by it : it has had no determinate meaning in their minds. Should any man ever find for it any determinate meaning, it will be this—that of the three branches between which, in this constitution, the aggregate powers of government are divided, it depends upon the will of each to prevent the two others from doing any thing —from giving effect to any proposed measure. How, by such arrangement, evil should be produced, is easy enough to say : for of this state of things one sure effect is—that whatsoever is in the judgment of any one of them contrary to its own sinister interest, will not be done ; on the other hand, notwithstanding the supposed security, whatsoever measure is by them all seen or supposed to be conducive to the aggregate interest of them all, will be carried into effect, how plainly soever it may be contrary to the universal interest of the people.   No abuse, in the preservation of which they have each an interest, will ever, so long as they can help it, be removed: no improvement in the prevention of which any one of them has an interest, will ever be made.

The fact is, that wherever on this occasion the word *balance* is employed, the sentence is mere nonsense. By the word *balance* in its original import, is meant a pair of scales.   In an arithmetical account, by an ellipsis to which, harsh as it is, custom has given its sanction, it is employed to signify that sum by which the aggregate of the sums that stand on one side of an account, exceeds the aggregate of the sums that stand on the other side of that same account.   To the idea which, in the sort of occasion in question, the word *balance* is employed to bring to view, this word corresponds not in any degree in either of these senses.   To accord with the sort of conception which, if any, it seems designed to convey, the word should be—not balance, but equipoise.   When two bodies are so connected that whenever the one is in motion, the other is in motion likewise, and *that* in such sort that in proportion as one rises the other falls, and yet at the moment in question no such motion has place, the two bodies may be said to be in *equipoise*; one

weighs exactly as much as the other. But of the
figure of speech here in question, the object is not to
present a clear view of the matter, but to prevent any
such view of it from being taken: to this purpose therefore, the nonsensical expression serves better than any
significant one. The ideas belonging to the subject
are thrown into confusion, the mind's eye in its endeavours to see into it is bewildered; and this is what
is wanted.

It is by a series of simultaneous operations that the
business of government is carried on: by a series of
actions:—action ceasing, the body politic, like the
body natural, is at an end. By a balance, if any
thing, is meant a pair of scales with a weight in each:
the scales being even, if the weights are uneven, that
in which is the heaviest weight begins to move; it
moves downward, and at the same time the other
scale with the weight in it moves upwards. All the
while this motion is going on no equipoise has place;
the two forces do not balance each other; if the wish
is that they should balance each other, then into the
scale which has in it the lighter weight must be added such other weight as shall make it exactly equal
to the heavier weight: or, what comes to the same
thing, a correspondent weight taken from that scale
which has in it the heavier weight.

The balance is now restored. The two scales hang
even; neither of the two forces preponderates over
the other. But with reference to the end in view, or

which ought to be in view—the use to be derived from the machine—what is the consequence?—All motion is at an end.

In the case in question, instead of two, as in a common pair of scales, there are three forces which are supposed, or said to be, antagonizing with one another. But were this all the difference, no conclusive objection to the metaphor could be derived from it; for, from one and the same fulcrum or fixed point you might have three scales hanging with weights in them, if there were any use in it: in the expression the image would be more complicated, but in substance it would be still the same.

Preeminently indeterminate, indistinct, and confused, on every occasion, is the language in which, to the purpose in question, application is made to this image of a balance; and on every occasion, when thus steadily looked into, it will be found to be neither better nor worse than so much nonsense. Nothing can it serve for the justification of: nothing can it serve for the explanation of.

The fallacy often assumes a more elaborate shape. " The constitution is composed of three forces, which, antagonizing with each other, cause the business of government to be carried on in a course which is different from the course in which it would be carried on if directed solely by any one, and is that which results from the joint influence of them all, each one of them contributing in the same proportion to the production of it."

Composition and resolution of forces. This image, though not so familiar as the other, is free from the particular absurdity which attaches upon the other : but upon the whole the matter will not be found much mended by it. In proportion as it is well conducted, the business of Government is uniformly carried on in a direction tending to a certain end,—the greatest happiness of the greatest number :—in proportion as they are well conducted, the operations of all the agents concerned, tend to that same end. In the case in question, here are three forces each tending to a certain end : take any one of these forces ; take the direction in which it acts; suppose that direction tending to the same exclusively legitimate end, and suppose it acting alone, undisturbed, and unopposed, the end will be obtained by it : add now another of these forces; suppose it acting exactly in the same direction, the same end will be attained with the same exactness, and attained so much the sooner : and so again if you add the third. But that second force—if the direction in which it acts be supposed to be ever so little different from that exclusively legitimate direction in which the first force acts, the greater the difference, the further will the aggregate or compound force be from attaining the exact position of that legitimate end.

But in the case in question how is it with the three forces? So far from their all tending to that end, the end they tend to is in each instance as opposite to that end as possible. True it is, that amongst these three several forces, that sort of relation really has place by which the sort of compromise in question is produced : a sort of direction which is not exactly the same as that which would be taken on the supposition that any one of the three acted alone, clear of the influence of both the others.   But with all this complication, what is the direction taken by the machine ?   Not that which carries it to the only legitimate end, but that which carries it to an end not very widely distant from the exact opposite one.

In plain language, here are two bodies of men, and one individual more powerful than the two bodies put together—say three powers—each pursuing its own interest, each interest a little different from each of the two others, and not only different from, but opposite to, that of the greatest number of the people :— of the substance of the people, each gets to itself and devours as much as it can.   Each of them, were it alone, would be able to get more of that substance, and accordingly would get more of that substance, than it does at present.   But in its endeavours to get that more, it would find itself counteracted by the two others : each therefore permits the two others to get their respective shares, and thus it is that harmony is preserved.

Balance of forces.—A case there is in which this metaphor, this image, may be employed with propriety : this is the case of international law and international relations.   Supposing it attainable, what is meant by a balance of forces or a balance of power is a legitimate object;—an object, the effectuation of which is beneficial to all the parties interested. What is that object? It is in one word *rest* ;—rest, the absence of all hostile motion, together with the absence of all coercion exercised by one of the parties over another: that rest which is the fruit of mutual and universal independence. Here then, as between nation and nation, that rest which is the result of well-balanced forces is peace and prosperity. But on the part of the several official authorities and persons by whose operations the business of Government in its several departments is carried on, is it prosperity that rest has for its consequence? No: on the contrary, of universal rest, in the forces of the body politic as in those of the body natural, the consequence is death. No action on the part of the officers of Government, no money collected in their hands,—no money, no subsistence,—no subsistence, no service,—no service, every thing falls to pieces, anarchy takes the place of government, government gives place to anarchy.

The metaphor of the balance, though so far from being applicable to the purpose in question, is in itself plain enough: it presents an image. The metaphor of the composition of forces is far from being so, —it presents not any image. To all but the comparatively few to whom the principles of mechanics, together with those principles of geometry that are associated with them, are thus far familiar, they present no conception at all: the conversion of the two tracts described by two bodies meeting with one another at an angle formed by the two sides of a parallelogram into the tract described by the diagonal of the parallelogram, is an operation never performed for any purpose of ordinary life, and incapable of being performed otherwise than by some elaborate mechanism, constructed for this and no other purpose.

When the metaphor here in question is employed, the three forces in question, the three powers in question, are, according to the description given of them, the power of the monarch, the power of the House of Lords, and the power of the people. Even according to this statement, no more than as to a third part of it would the interest of the people be promoted : as to two-thirds it would be sacrificed. For example:— out of every 300*l.* raised upon the whole people, one hundred would be raised for the sake and applied to the use of the whole people ;—the two other thirds for the sake and to the use of the two confederative powers, to wit the monarch and the House of Lords.

Not very advantageous to the majority of the people, not very eminently conducive to good government, would be this state of things ; in a prodigious degree, however, more conducive would it be than is the real state of things. For, in the respect in question, what is this real state of things? The power described as above by the name of the power of the people, is, instead of being the power of the people, the power of the monarch and the power of the House of Lords, together with that of the rest of the aristocracy under that other name.

5. *Glorious Revolution.*

This is a Whig's cry, as often as it is a time to look bold, and make the people believe that he had rather be hanged than not stand by them. What? a revolution for the people? No: but, what is so much better, a revolution for the Whigs,—a revolution of 1688. There is your revolution: the only one that should ever be thought of without horror,—a revolution for discarding kings? No: only a revolution for changing them. There would be some use in changing them, there would be something to be got by it. When their forefathers of 1688 changed James for William and Mary, William got a good slice of the cake, and they got the rest among them. If, instead of being changed, kings were discarded, what would the Whigs get by it? They would get nothing;—they would lose not a little: they would lose their seats, unless they really sat and did the business they were sent to do, and then they would lose their ease.

The real uses of this revolution were the putting an end to the tyranny, political and religious, of the Stuarts:—the political, governing without parliament, and forcing the people to pay taxes without even so much as the show of consenting to them by deputies chosen by themselves:—the religious, forcing men to join in a system of religion which they believed not to be true.

But the deficiencies of the revolution were, leaving the power of governing, and in particular that of taxing, in the hands of men whose interest it was to make the amount of the taxes excessive, and to exercise misrule to a great extent in a great variety of other ways.

So far as by security given to all, and thence, by check put to the power of the crown, the particular interest of the aristocratical leaders in the revolution promised to be served, such security was established, such check was applied. But where security could not be afforded to the whole community without trenching on the power of the ruling few, there it was denied. Freedom of election, as against the despotic power of the monarch, was established;—freedom of election, as against the disguised despotism of the aristocracy, Tories and Whigs together, remained excluded.

CHAPTER IV.

Allegorical Idols.

Ad imaginationem.

Exposition.

THE use of this fallacy is the securing to persons
in office, respect independent of good behaviour. This
is in truth only a modification of the fallacy of vague
generalities, exposed in the preceding chapter. It consists in substituting for men's proper official denomination the name of some fictitious entity, to whom,
by customary language, and thence opinion, the attribute of excellence has been attached.

Examples 1. *Government;* for members of the
governing body. 2. *The church;* for churchmen.
3. *The law;* for lawyers. The advantage is, obtaining for them more respect than might be bestowed on
the class under its proper name.

Exposure.

I. *Government.*—In its proper sense, in which it
designates the set of operations, it is true, and universally acknowledged, that every thing valuable to
man depends upon it: security against evil in all
shapes, from external adversaries as well as domestic.
II. *Law:* execution of the *law.*—By this it is that
men receive whatsoever protection they receive against
domestic adversaries and disturbers of their peace. By

*government,—law—the law,—*are therefore brought
to view the naturalest and worthiest objects of respect
and attachment within the sphere of a man's observance : and for conciseness and ornament (not to
speak of deception) the corresponding fictitious entities are feigned, and represented as constantly occupied in the performance of the above-mentioned all-
preserving operations.   As to the real persons so occupied, if they were presented in their proper character, whether collectively or individually, they would
appear clothed in their real qualities, good and bad
together.   But, as presented by means of this contrivance, they are, decked out in all their good and acceptable qualities, divested of all their bad and unac-
ceptable ones.   Under the name of the god Æscula-
pius, Alexander the impostor, his self-constituted high
priest, received to his own use the homage and offer-
ings addressed to his god.   Acquired, as it is believed,
comparatively within late years, this word *government*
has obtained a latitude of import in a peculiar degree
adapted to the sinister purpose here in question.   From
abstract, the signification has become, as the phrase
is, *concrete.*   From the system, in all its parts taken
together, it has been employed to denote the whole
assemblage of the individuals employed in the carrying on of the system : of the individuals who, for the
time being, happen to be members of the official establishment, and of these more particularly, and even
exclusively, such of them as are members of the ad-

ministrative branch of that establishment. For the
designation either of the branch of the system or of
the members that belong to it, the language had already
furnished the word *Administration.* But the word
Administration would not have suited the purpose of
this fallacy: accordingly, by those who feel themselves
to have an interest in the turning it to account, to the
proper word Administration the too ample and thence
improper word *Government* has been, probably by a
mixture of design and accident, commonly substituted.

This impropriety of speech being thus happily and
successfully established, the fruits of it are gathered
in every day. Point out an abuse,—point to this or
that individual deriving a profit from the abuse, up
comes the cry, " You are an enemy to government!"
then, with a little news in advance, " Your endeavour
is to destroy government !" thus you are a Jacobin, an
anarchist, and so forth : and the greater the pains you
take for causing government to fulfill, to the greatest
perfection, the professed ends of its institution, the
greater the pains taken to persuade those who wish,
or are content, to be deceived, that you wish and endeavour to destroy it.

III. *Church.*—This is a word particularly well
adapted to the purpose of this fallacy: to the elements
of confusion shared by it with *government* and *law,* it
adds divers, proper to itself. The significations indifferently attachable to the word *Church,* are, 1. Place
of worship; 2. Inferior officers engaged by Govern-

ment to take a leading part in the ceremonies of worship [a].   3. All the people considered as worshippers. 4. The superior officers of Government by whom the inferior, as above, are engaged and managed. 5. The rules and customs respecting those ceremonies.

The use of this fallacy to churchmen, is the giving and securing to them a share of coercive power; their sole public use and even original destination being the serving the people in the capacity of instructors,—instructing them in a branch of learning, now more thoroughly learnt without than from them [b].   In the phrase "*church* and *state*," *churchmen* are represented as superior to all *non-churchmen*.   By "*church* and *king*," churchmen are represented as superior to the king.   Fox and Norfolk were struck off the list of privy counsellors for drinking " The sovereignty of the people;" the reduction would be greater were all struck off who have ever drank " Church and king." According to Bishop Warburton's Alliance, the people in the character of the church, meeting with all themselves in the character of the state, agreed to invest the expounders of the sacred volume with a large share of the sovereignty.   Against this system, the lawyers, their only rivals, were estopped from pleading its seditiousness in bar.   In Catholic countries, the churchmen who compose holy mother church possess one beautiful female, by whom the people are governed in

---

[a] Articles 19, 20.

[b] *ex. gr.* from unordained Methodists, &c. and Quakers.

the field of spiritual law, within which has been in-
closed as much as possible of profane law.   By Protestants, on holy mother church the title of Whore of
Babylon has been conferred : they recognise no holy
mother church.   But in England, churchmen, a large
portion of them, compose two *Almæ Matres Acade-
miæ*—kind Mother Academies or Universities.   By
ingenuity such as this, out of "lubberly post-masters'
boys" in any number, one "sweet Mrs. Anne Page"
is composed, fit to be decked out in elements of amia-
bility to any extent.   The object and fruit of this ingenuity is the affording protection to all abuses and
imperfections attached to this part of the official establishment.   Church being so excellent a being, none
but a monster can be an *enemy*, a *foe*, to her.   *Monster*, i. e. anarchist, Jacobin, leveller, &c.   To every
question having reform or improvement in view as to
this part of the official establishment, the answer is
one and the same—"You are an enemy to the church."
For instance, among others, to such questions as follow : 1. What does this part of the official establishment do, but read or give further explanation to one
book, of which more explanation has been given already than the longest life would suffice to hear?
2. Does not this suppose a people incapable of being
taught to read?   3. Would it not be more read if
each of them, being able to read, had it constantly by
him to read all through, than by their being at liberty
some of them to go miles to hear small parts of it?—
Suppose it admitted, that by the addition of other ser-

vices conducive to good morals and good government, business for offices not much inferior to the existing ecclesiastical offices might be found,—then go on and ask, —1. As to the connexion between reward and service, do not the same rules apply to these as to profane offices? 2. Pay unconditioned for service, is it more effectual in producing service here than there? 3. *Here* more than *there*, can a man serve in a place without being there? 4. Here, as there, is not a man's relish for the business proved the greater, the smaller the factitious reward he is content to receive for doing it? 5. The stronger such his relish, is not his service likely to be the better? 6. Over and above what, if any thing, is necessary to engage him to render the service, does not every penny contribute to turn him aside to other and expensive occupations, by furnishing him with the means? 7. In Scotland, where there is less pay, is not *residence* more general, and clerical *service* more abundant and efficient?

Answer,—Enemy!—and, if English-bred,—Apostate!

1. In Scotland, does any evil arise from the non-existence of bishops? 2. In the House of Lords, any good? 3. Is not non-attendance there more general than even non-residence elsewhere? 4. *In judiciali*, does any bishop ever attend, who is not laid hold of after reading prayers? 5. *In legislatura*, ever, except where personal interest wears the mask of gratitude? 6. Such non-attendance, is it not felt rather as a relief than as a grievance?

Answer,—" Enemy to the church !"

1. In Ireland, what is the use of Protestant priests to Catholics, who will neither hear nor see them, to whom they are known but as plunderers? 2. By such exemption from service, is not *value* of preferment increased? 3. By patrons, as by incumbents, are not bishopricks thus estimated? 4. Is it not there a maxim, that service and pay should be kept in separate hands? 5. In eyes not less religious than gracious, is not the value of religion inversely as the labour, as well as directly as the profit? 6. Is not this estimate the root of those scruples, by which oaths imposed to protect Protestantism from being oppressed are employed in securing to it the pleasure of oppressing?

Answer,—" Enemy to the church !"

CHAPTER V.

Sweeping Classifications.

Ad judicium.

Exposition.

THE device of those who employ in the way of fallacy, sweeping classifications, is that of ascribing to an individual object (person or thing), any properties of another, only because the object in question is ranked in the class with that other, by being designated by the same name.

In its nature, this fallacy is equally applicable to undeserved eulogy as to undeserved censure; but it is more frequently applied to the purpose of censure, its efficiency being greater in that direction.

Exposure.

Example 1.—*Kings ;—Crimes of Kings.*

In the heat of the French revolution, when the lot of Louis XVI. was standing between life and death, among the means employed for bringing about the catastrophe that ensued, was the publication of a multitude of inflammatory pamphlets, one of which had for its title " The Crimes of Kings."

Kings being men, and all men standing exposed to those temptations by which some of them are led into crimes, matter could not be wanting for a book so entitled : and if there are some crimes to the temptation of which men thus elevated stand less exposed than the inferior orders, there are other crimes to which, perhaps, that elevation renders them but the more prone.

But of the man by whom on that occasion a book with such a title was published, the object, it is but too probable, was to compose out of it this argument: criminals ought to be punished; kings are criminals, and Louis is a king: therefore Louis ought to be punished.

Example 2.—*Catholics ; cruelties of Catholics.*

Not long ago, in the course, and for the purpose of the controversy on the question whether that part of the community which is composed of persons of the Catholic persuasion, ought, or ought not, to be kept any longer in a state of degradation under the predominant sect, a book made its appearance under the title of " Cruelties of the Catholics."

Of any such complete success, as the consigning in a body to the fate in which that Catholic king was, with so many of his nearest connections, involved, all such British subjects as participate with him in that odious name, there could not be much hope : but whatsoever could, by the species of fallacy here in question, be done towards the promoting of it, was done by that publication. The object of it was to keep them still debarred from whatsoever relief remains yet to be administered to the oppressions under which they labour : either it had this object, or it had none.

To the complexion of this argument, and of the mind that could bring it forward, justice will not be done, unless an adequate conception be formed of the practical consequences to which, if to any thing, it leads.

Of the Catholics of the present and of all future time, whatsoever be the character, the cruelties and other enormities committed by persons who in former times were called by the same indefinitely comprehensive name, will still remain what they were. Whatsoever harsh treatment, therefore, this argument warrants the bestowing on these their namesakes at the present time, the same harsh treatment will, from the same argument, continue to receive the same justification so long as there remains one individual who, consistently with truth, is capable of being characterized by the same name.

Be they what they may, the barbarities of the Catholics of those times had their limits : but of this abhorrer of Catholic barbarities, the barbarity has, in respect of the number of intended victims, no limits other than those of time.

Of the man who, to put an end to the cruelties of kings, did what depended upon him towards extirpating the class of kings, the barbarity, so far as regarded this object, was, comparatively speaking, confined within a very narrow range. All Europe would not have sufficed to supply his scaffold with a dozen victims. But after crushing as many millions of the vermin whom his piety and his charity marked out for sacrifice, the zeal of the abhorrer of Catholic cruelties would have been in the condition of the tiger whom in the plains of Southern Africa a traveller depicted to us as lying breathless with fatigue amidst a flock of antelopes.

In the same injurious device the painter of the crimes of kings might, by a no less conclusive argument, have proved the necessity of crushing the English form of the Protestant religion, and consigning to the fate of Louis XVI. the present head of it.

By order of King James I. two men, whose misfortune it was not to be able to form in relation to some inexplicable points of technical theology the same conception that was entertained, or professed to be entertained, by the royal ruler and instructor of his people, were burnt alive[a]. George IV. not only bears in common with James I. the two different denominations, viz. Protestant of the Church of England, and King of Great Britain; but, as far as marriage can be depended on for proof of filiation, is actually of the same blood and lineage with that royal and triumphant champion of local orthodoxy.

If, indeed, in the authentic and generally received doctrines of the religion in question, there were any thing that compelled its professors to burn or other-

---

[a] Consult Hume, Tindal, Harris, Henry.

wise to destroy or ill-treat all or any of those that differed from them, and if by any recent overt-act an
adherence to those dissocial doctrines had appeared
in practice, in such case the adherence to such dissocial doctrines would afford a just ground for whatsoever measures of security were deemed necessary to
guard other men from the effect of such doctrines and
such practice.

But by no doctrines of their religion are Catholics
compelled to burn or otherwise ill-treat those who differ from them, any more than by the doctrines of the
Church of England James I. was compelled to burn
those poor Anabaptists.

If from analogy any sincere and instructive use had
on this occasion been intended to be derived from different countries professing the same persuasion, in
these our times a much more instructive lesson would
be afforded than any that could be derived from even
the same country at such different times.

If in Ireland, where three-fourths or more of the
population is composed of Catholics, no ill-treatment
has within the memory of man been bestowed by Catholics, as such, upon Protestants, as such; while in
the same country so much ill-treatment has on other
accounts been bestowed by each of these persuasions
upon the other, it is, it may be said, because the power
of doing so with impunity is not in their hands.

But in countries where the Catholic religion is the
predominant religion, and in which at the same time,
as in our islands, barbarity on the score of heresy was

by Catholics exercised according to law, and in the countries in which the exercise of those barbarities was at those times most conspicuous ; of no such barbarities has any instance occurred for a long course of years[a].

---

[a] Even in Spain I have been assured, if I may depend upon an assurance given me by persons fully informed, and of the most respectable character, no instance of a capital execution for any offence against religion has occurred within these 22 or 23 years.

In the capital of Mexico, if I may believe a gentleman of distinction in our own country, by whom the capital of that kingdom was lately visited, he was by the Grand Inquisitor himself conducted into every apartment of the prison of the Inquisition for the purpose of his being assured by ocular demonstration of the non-existence of any person in the state of a prisoner within the walls.

CHAPTER VI.

Sham Distinctions.

Ad judicium.

Exposition.

Of the device here in view the nature may be explained by the following direction for the use of it.

When any existing state of things has too much evil in it to be defensible *in toto*, or proposals for amendment are too plainly necessary to be rejectible *in toto*, the evil and the good being nominally distinguished from each other by two corresponding and opposite terms, eulogistic and dyslogistic, but in such sort that to the nominal line of distinction thus drawn, there corresponds not any determinate real difference, declare your approbation of the good by its eulogistic name, and thus reserve to yourself the advantage of opposing it without reproach by its dyslogistic name, and so *vice versa* declare your disapprobation of the evil, &c.

Exposure.

Example 1.—*Liberty and Licentiousness of the Press.*

Take for example the case of the Press.

The press (including under this denomination every instrument employed or employable for the purpose of giving diffusion to the matter of human discourse by visible signs)—the press has two distinguishable uses, viz. moral and political: moral, consisting in whatsoever check it may be capable of opposing to misconduct in private life: political, in whatsoever check it may be capable of opposing to misconduct in public life, that is, on the part of public men—men actually employed, or aspiring to be employed, in any situation in the public service: opposing, viz. by pointing on the persons to whom such misconduct is respectively imputable, a portion more or less considerable of disapprobation and consequent ill-will on the part of the public at large—a portion more or less considerable according to the nature of the case.

If to such misconduct there be no such check at all opposed, as that which it is the nature of the press to apply, the consequence is, that of such misconduct whatsoever is not included in the prohibitions and eventual punishment provided by law, will range uncontrolled: in which case, so far as concerns the political effect of such exemption from control, the result is power uncontroulable, arbitrary despotism, in the hands whatsoever they are, in which the powers and functions of Government happen to be reposed: and, moreover, in the instance of such misconduct as is included in that system of prohibition and eventual punishment, the control will be without effect in so far as by delay, vexation, and expense, natural or factitious, the individual who would be led to call for the application, is prevented from making such demand.

At the same time, on the other hand, the use of the

press cannot be altogether free, but that on pretence of giving indication of misconduct that has actually taken place, supposed misconduct that never did actually take place, will to this or that individual be imputed.

In so far as the imputation thus conveyed happens to be false, the effects of the liberty in question will, so far as concerns any individual person thus unjustly accused, be of the *evil* cast, and by whomsoever they are understood so to be, the dyslogistic appellation *licentiousness* will naturally be applied.

Here then comes the dilemma,—the two evils between which a choice must absolutely be made. Leave to the press its perfect liberty, along with the just imputations, which alone are the useful ones, will come, and in an unlimited proportion, unjust imputations, from which, in so far as they are unjust, evil is *liable* to arise.

But to him whose wish it really is that good morals and good government should prevail, the choice need not be so difficult as at first sight it may seem to be.

Let all just imputations be buried in utter silence ; what you are sure of is, that misconduct in every part of the field of action, moral and political, private and public, will range without controul—free from all that sort of controul which can be applied by the press, and not by any thing else.

On the other hand, let all unjust imputations find through this channel an unobstructed course, still, of the evil—the personal suffering threatened by such

T

infliction—there is neither certainty, nor in general
any near approach to it. Open to accusation, that
same channel is not less open to defence[a]. He therefore who has truth on his side, will have on his side
all that advantage which it is in the nature of truth to
give.

That advantage, is it an inconsiderable one?—On
the contrary supposition is founded, whatsoever is
done in the reception and collection of judicial evidence—whatsoever is intended by the exercise of judicial authority—by the administration of whatsoever
goes by the name of justice.

Meantime, if any arrangements there be by which
the door may be shut against unjust imputations, without incurring to an equal amount that sort of evil
which is liable to result from the exclusion of just
ones, so much the better.

But unless and until such arrangements shall have
been devised and carried into effect, the tendency and
effect of all restrictions having for their object the
abridging of the liberty of the press, cannot but be
evil on the whole.

To shut the door against such imputations as are
either unjust or useless, leaving it at the same time
open to such as are at the same time just and useful,
would require a precise, a determinate, a correct and
complete definition of the appellative, whatsoever it

---

[a] If it by accident be not so, this constitutes a different and distinct
evil, for which is required a different and distinct remedy.

be, by which the abuse—the improper use—the supposed preponderantly pernicious use—of the press is endeavoured to be brought to view.

To establish this definition belongs to those, and to those alone, in whose hands the supreme power of the State is vested.

Of this appellative no such definition has ever yet been given—of this appellative no such definition can reasonably be expected at the hands of any person so situated, since, by the establishing of such definition, their power would be curtailed, their interest prejudiced.

While this necessary definition remains unestablished, there remains with them the faculty of giving continuance and increase to the several points of abuse and misgovernment by which their interest in its several shapes is advanced.

Till that definition is given, the *licentiousness* of the press is every disclosure by which any abuse from the practice of which they draw any advantage, is brought to light and exposed to shame :—whatsoever disclosure it is, or is supposed to be, their interest to prevent.

The *liberty* of the press is such disclosure, and such only, from which no such inconvenience is apprehended.

No such definition can be given but at their expense :—at the expense of their arbitrary power,—of their power of misconduct in the exercise of the functions of Government,—at the expense of their power of misgovernment,—of their power of sacrificing the public interest to their own private interest.

Should that line have ever been drawn, then it is that licentiousness may be opposed without opposing liberty: while that line remains undrawn, opposing licentiousness is opposing liberty.

Thus much being understood, in what consists the device here in question? It consists in employing the sham approbation given to the species of liberty here in question under the name of *liberty*, as a mask or cloak to the real opposition given to it under the name of *licentiousness*.

It is in the licentiousness of the press that the Judge pretends to see the downfall of that Government, the corruption of which he is upholding by inflicting on all within his reach those punishments which by his predecessors have been provided for the suppression of all disclosures, by means of which the abuses which he profits by might be checked.

Example 2.—*Reform, temperate and intemperate.*

For the designation of the species or degree of political reform which, by him who speaks of it, is meant to be represented as excessive or pernicious, the language affords no such single-worded appellative as in the case of *liberty*:—the liberty of the press. For making the nominal and pretended real distinction, and marking out on the object of avowed reprobation the pernicious or excessive species or degree, recourse must therefore be had to epithets or adjuncts: such for instance as violent, intemperate, outrageous, theoretical, speculative, and so forth.

If, with the benefit of the subterfuge afforded by any of these dyslogistic epithets, a man indulges himself in the practice of reprobating reform in terms thus vague and comprehensive, and without designating by any more particular and determinate word the species or degree of reform to which he means to confine his reprobation, or the specific objections he may have to urge, you may in general venture to conclude it is not to any determinate species or degree that his real disapprobation and intended opposition confines itself, but that it extends itself to every species or degree of reform which, according to his expectation, would be efficient: that is, by which any of the existing abuses would find a corrective.

For, between all abuses whatsoever, there exists that connexion,—between all persons who see each of them any one abuse in which an advantage results to himself, there exists in point of interest that close and sufficiently understood connexion, of which intimation has been given already. To no one abuse can correction be administered without endangering the existence of every other.

If, then, with this inward determination not to suffer, so far as depends upon himself, the adoption of any reform which he is able to prevent, it should seem to him necessary or advisable to put on for a cover, the profession or appearance of a desire to contribute to such reform,—in pursuance of the device or

fallacy here in question he will represent that which
goes by the name of reform as distinguishable into two
species ; one of them a fit subject for approbation, the
other for disapprobation. That which he thus professes to have marked for approbation he will accordingly, for the expression of such approbation, characterize by some adjunct of the *eulogistic* cast, such as
moderate, for example, or temperate, or practical, or
practicable.

To the other of these nominally distinct species he
will, at the same time, attach some adjunct of the
dyslogistic cast, such as violent, intemperate, extravagant, outrageous, theoretical, speculative, and so forth.

Thus, then, in profession and to appearance, there
are in his conception of the matter two distinct and
opposite species of reform, to one of which his approbation, to the other his disapprobation, is attached.
But the species to which his approbation is attached
is an *empty* species,—a species in which no individual
is, or is intended to be, contained.

The species to which his disapprobation is attached
is, on the contrary, a crowded species, a receptacle in
which the whole contents of the *genus*—of the genus
reform—are intended to be included.

CHAPTER VII.

Popular Corruption.

Ad superbiam.

Exposition.

THE instrument of deception, of which the argument here in question is composed, may be thus expressed :—The source of corruption is in the minds of the people; so rank and extensively seated is that corruption, that no political reform can ever have any effect in removing it [a].

Exposure.

This fallacy consists in giving to the word *corruption*, when applied to the people, a sense altogether

---

[a] This was an argument brought forward against parliamentary reform by William Windham in the House of Commons, and by him insisted on with great emphasis. This man was among the disciples, imitators of, and co-operators with, Edmund Burke—that Edmund Burke with whom the subject many were the swinish multitude :— swinish in nature, and apt therefore to receive the treatment which is apt to be given to swine. In private life, that is, in their dealings with those who were immediately about them, at any rate such of them as were of their own class, many of these men, many of these haters and calumniators of mankind at large, are not unamiable ; but, seduced by that sinister interest which is possessed by them in common, they encourage in one another the antisocial affection in the case where it operates upon the most extensive scale. If, while thus encouraging himself in the hating and contemning the people, a man of this cast finds himself hated by them, the fault is surely more in him than them ; and, whatever it may happen to him to suffer from it, he has himself to thank for it.

indeterminate,—a sense in and by which all that is
distinctly expressed is the disaffection of the speaker
as towards the persons spoken of, imputing to them
a bad moral character or cast of mind, but without
any intimation given of the particular nature of it.

It is the result of a thick confusion of ideas, whether
sincere, or affected for the purpose.

In the case of a parliamentary election, each elector acts as a trustee for himself and for all the other
members of the community in the exercise of the
branch of political power here in question. If, by the
manner in which his vote is received from him, he is
precluded (as by ballot) from the possibility of promoting his own particular interest to the prejudice of
the remainder of the universal interest, the only interest of his which he can entertain a prospect of promoting by such his vote is his share of the universal
interest: and for doing this, he sees before him no
other possible means than the contributing to place
the share of power attached to the seat in question in
the hands of that candidate who is likely to render
most service to the universal interest.

Now, how inconsiderable soever may be in his eyes
this his share in the universal interest, still it will be
sufficient to turn the scale where there is nothing in
the opposite scale: and, by the supposition, the empti-
ness of the opposite scale has been secured in the
mode of election by ballot, where the secrecy thereby
endeavoured at is accomplished, as to so complete a
certainty it may be. If then, to continue the allusion,

the value of his share in the universal interest, in his
eyes, is such as to overcome the love of ease,—the
aversion to labour,—he will repair to the place, and
give his vote,—to that candidate who, in his eyes, is
likely to do most service to the universal interest: if
it be not sufficient to overcome that resisting force, he
will then forbear to give his vote ; and though he will
do no good to the universal interest, he will do no
harm to it.

Thus it is that, under an apposite system of election procedure, supposing them in the account of
self-regarding prudence equal, the least benevolent set
of men will, on this occasion, render as much service
to the universal interest as the most benevolent : the
least benevolent, if that be what is meant by the *most*
corrupt ; and if that is not meant, nothing which is to
the purpose, nor in short any thing which is determinate, is meant.

On the other hand, in so far as the system of election is so ordered, that by the manner in which he
gives his vote a man is enabled to promote his own
separate interest, what is sufficiently notorious is, that
no ordinary portion of benevolence in the shape of
public spirit will suffice to prevent the breach of trust
in question from being committed.

In the case, therefore, of the subject many, to whom
exclusively it was applied, the word *corruption* has no
determinate and intelligible application.   But to the
class of the ruling few, it has a perfectly intelligible

application;—application in a sense in which the truth of it is as notorious as the existence of the sun at noonday. Pretending to be all of them chosen by the subject many,—chosen in fact, a very small proportion of them in that manner,—the rest by one another,—they act in the character of trustees for the subject many, bound to support the interest of the subject many: instead of so doing, being with money exacted from the subject *many* bribed by one another acting under the ruling *one*, they act in constant breach of such their trust, serving in all things their own particular and sinister interests at the expense and to the sacrifice of that interest of the subject many, which, together with that of the ruling few, composes and constitutes the universal interest. *Corrupt, corruption, corruptors, corruptionist,* applied to conduct such as hath been just described,—the meaning given to these terms wants assuredly nothing of being sufficiently intelligible.

A circumstance that renders this fallacy in a peculiar degree insidious and dangerous, is a sort of obscure reference made by it to certain religious notions: to the doctrine of original sin as delivered in the compendium of Church of England faith, termed the 39 articles.

Into that doctrine, considered in a religious point of view, it is not necessary on this occasion to make any inquiry. The field here in question is the field of politics; and applied to this field the fallacy in question seeks to lay the axe to the root of all government. It applies not only to *this*, but to all other remedies against that preponderance of self-regarding over social interest and affection, which is essential to man's existence, but which for the creation and preservation of political society, and thence for his well-being in it, requires to be checked—checked by a force formed within itself. It goes to the exclusion of all laws, and in particular of all penal laws; for if, for remedy to what is amiss, nothing is to be attempted by arrangements which, such as those relative to the principle and mode of election as applied to rulers, bring with them no punishment,—no infliction,—how much less should the accomplishment of any such object be attempted by means so expensive and afflictive as those applied by penal laws!

By the employment given to this fallacy, the employer of it afforded himself a double gratification;—he afforded an immediate gratification to his own antisocial pride and insolence, while he afforded to his argument a promise of efficiency by the food it supplied to the same appetite in the breasts of his auditors, bound to him, as he saw them to be, by a community of sinister interest.

Out of the very sink of immorality was this fallacy drawn: a sentiment of hatred and contempt, of which not only all the man's fellow-countrymen were the declared, but all mankind in at least equal degree were the naturally supposable, object. "So bad are they in themselves, no matter how badly they are treated : they cannot be treated worse than they deserve : Of a bad bargain (says the proverb) make the best; of so bad a crew, let us make the best for ourselves : no matter what they suffer, be it what it may, they deserve it." If Nero had thought it worth his while to look out for a justification, he could not have found a more apt one than this : an argument which, while it harmonized so entirely with the worst passions of the worst men, screened its true nature in some measure from the observation of better men, by the cloud of confusion in which it wrapped itself.

In regard to corruption and uncorruption, or, to speak less ambiguously, in regard to vice and virtue, how then stands the plain and real truth? That in the ruling few there is most vice and corruption, because in their hands has been the power of serving their own private and sinister interest at the expense of the universal interest : and in so doing they have, in the design and with the effect of making instruments of one another for the accomplishment of that perpetual object, been the disseminators of vice and corruption :—That in the subject many, there has been least of vice and corruption, because they have not been in so large a degree partakers in that sinister interest, and have thus been left free to pursue the track pointed out to them, partly by men who have found a personal interest in giving to their conduct a universally beneficial direction,—partly by discerning and uncorrupted men, who, lovers of their country and mankind, have not been in the way of having that generous affection overpowered in their breasts by any particular self-regarding interest.

Nearly akin to the cry of popular corruption is language commonly used to the following effect :—" Instead of reforming others, instead of reforming your betters, instead of reforming the state, the constitution, the church, every thing that is most excellent,—let each man reform himself—let him look at home, he will find there enough to do, and what is in his power, without looking abroad and aiming at what is out of his power, &c. &c."

Language to this effect may at all times be heard from anti-reformists, always, as the tone of it manifests, accompanied with an air of triumph,—the triumph of superior wisdom over shallow and presumptuous arrogance.

One feature which helps to distinguish it from the cry of popular corruption, is the tacit assumption that, between the operation condemned and the operation recommended, incompatibility has place: than which, when once brought clearly to view, nothing, it will be seen, can be more groundless.

Certain it is, that if every man's time and labour is exclusively employed in the correcting of his own personal imperfections, no part of it will be employed in the endeavour to correct the imperfections and abuses which have place in the Government, and thus the mass of those imperfections and abuses will go on, never diminishing, but perpetually increasing with the torments of those who suffer by them, and the comforts of those who profit by them : which is exactly what is wanted.

CHAPTER VIII.

Observations on the seven preceding Fallacies.

In the seven preceding fallacies, and in others of a similar nature, the device resorted to is uniformly the same, and consists in entirely avoiding the question in debate, by substituting general and ambiguous terms in the place of clear and particular appellatives.

In other fallacies the argument advanced is generally irrelevant, but argument of some kind they do contain. In these, argument there is none; *Sunt verba et voces prætereàque nihil.*

To find the only word that will suit his purpose, the defender of corruption is obliged to make an ascent in the scale of generalization, to soar into the region of vague generalities, till he comes to a word by the extensiveness of whose import he is enabled, so by confounding language, to confound conceptions, as without general and immediate fear of detection to defend with a chance of success an object of the defence of which there would under its proper and peculiar name be no hope.

When of two terms, viz. a generic term, and a specific term included under it, the specific term alone is proper, *i. e.* the proposition into the composition of which it enters, true; the generic term, if substituted to it, is ambiguous, and of the ambiguity, if the effect of it is not perceived, the consequence is error and deception.

Opposite to this *aërial* mode of contestation, is the mode already known and designated by the appellation of *close reasoning*.

In proportion as a man's mode of reasoning is close, (always supposing his intention honest,) for the designation of every object which he has occasion to bring to view, he employs in preference the most particular expression that he can find: that which is best adapted to the purpose of bringing to view every thing which is its object to bring to view, as clear as possible from every thing which the purpose does not require to be brought, and which in consequence it is his endeavour to avoid bringing, to view.

In proportion as a man is desirous of contributing on every occasion to the welfare of the community, and at the same time skilled in the means that most directly and certainly lead to the attainment of that end, he will, on the occasion of the language employed by him in the designation of each measure, look out for that plan of nomenclature and classification by which the degree and mode of its conduciveness or repugnancy to that end may be the more easily and correctly judged of.

Thus, in regard to offences—acts which on account of their adverseness to the general welfare are objects meet for discouragement,—for prohibition, and in case of necessity, for punishment,—not content with the employing for the designation of each such act in

particular, that mode of expression by which every individual act partaking of the common nature indicated
by the generic term may be brought to view, to the
exclusion of every act not partaking of that common
nature, he will, for the designation of the relation it
bears to other offences, and of the place which it occupies in the aggregate assemblage of these obnoxious
acts, find for it and assign to it some such more general and extensive appellation as shall give intimation
of the *mode* in which the wound given by it to the general welfare is perceptible.

1. Offences against individuals other than a man's
self, and those, assignable individuals. 2. Against a
man's self. 3. Against this or that particular class
of the community. 4. Against the whole community
without distinction.

In the case of individuals; offences against person,
against reputation, against property, against condition
in life ;—and so on through the other classes above
designated[a].

For the opposite reason, in proportion as without
regard to, and to the sacrifice of, the general welfare,
a man is desirous of promoting his own personal or
any other private interest, he will on the occasion of
the language employed in the designation of each
measure look out for that plan of nomenclature and
classification, by which the real tendency of the mea-

---

[a] See *Traités de Législation*, tom. i. p. 172.   *Classification des délits.*

U

sure to which he proposes to give birth or support, shall
be as effectually masked as possible :—rendered as difficult as possible to be comprehended and judged of.

In the English law under the principle of arrangement which till comparatively of late years was the
only one, and which is still the predominant one, such
were the groupes into which, by the classical denominations employed, they were huddled together, that by
those denominations not any the slightest intimation
was given of the nature and mischief of the offences
respectively contained under them.   Treasons, felonies unclergyable, felonies clergyable, premunires,
misdemeanors.

By the four first of these five denominations what is
designated, is,—not the offence itself,—but the treatment given to the offender in respect of it in the way
of punishment: by the other denomination not so
much as even that:—only that the act is treated on
the footing of an offence, and on that score made punishable : it is the miscellaneous class, the contents
of which are composed of all such offences as are not
comprised under any of the others.

To what cause can a scheme of arrangement so incompatible with clear conception and useful instruction be ascribed?

Its creation may be traced to one source :—its continuance to another.   For its creation, (such is its antiquity,) the weakness of the public intellect, presents
an adequate cause.   Of treason and felony,—terms

imported at the Norman conquest with the rest of the nomenclature of the feudal system,—the origin is lost in the darkness of primæval barbarism: religion, a perversion of the Christian religion, gave birth after a hard and long labour to the distinction between clergyable and unclergyable. Religion by a further perversion gave birth to *premunires* in the reign of Edward III.

To the designs of those whose interest it is that misrule in all its shapes should be perpetuated, and thence, that useful information, by which it might be put to shame, and in time to flight, should as long as possible be excluded, nothing could be more serviceable than this primæval imbecility. Under these denominations in general, and in particular under felony, acts of any description are capable of being ranked with equal propriety, or rather with equal absence of impropriety : acts of any description whatsoever, and consequently acts altogether pure from any of those mischievous consequences from which alone any sufficient warrant for subjecting the agents to punishment, can be found ; and offences thus clear of every really mischievous quality, have accordingly been created, and still continue in existence, in convenient abundance.

By this contrivance the open tyranny of the lawyer-led legislator, and the covert tyranny of the law-making judge are placed at the most perfect ease. The keenest eye cannot descry the felonies destined to be created by the touch of the sceptre upon the pattern

of the old : the liveliest imagination cannot pourtray
to itself the innoxious acts destined to be fashioned or
swollen into felonies.

Analogous to this ancient English system—correspondent and analogous both as to the effect itself,
and as to its cause, is the system lately brought out
by the legislators of France and their forced imitators in Germany. *Faute, contravention, délit, crime,*
classes, rising one above another in a climax of se-
verity,—all of them designative how indeterminately
soever, rather of the treatment to which at the hands
of the judge, the agent is subjected, than of the sort of
act for which he is subjected to that treatment, much
less of the ground, or reason, on which (regard being
had to the quality and quantity of mischief) it is
thought fit he shall be so dealt with.

Lawyer-craft in alliance with political tyranny may
be marked out as the source of this confusion in the
English case; lawyer-craft in subjection to political
tyranny in the French case.

In England it is the interest of the man of law that
the rule of action should be, and continue, in a state
of as general uncertainty and incognoscibility as possible : that on condition of pronouncing on each occasion a portion of the flash language adapted to that
purpose he may in his state of law-adviser and advocate, be master of men's purses ; in his state of judge,—
of purse, reputation, condition in life, and life itself,
to as complete a degree and with as little odium and
suspicion as possible. This is the state of things

which it always has been, and will be his interest to perpetuate : and this is the state of things which hitherto it has been in his power to continue, and which accordingly does to this day continue in existence.

In France where the man of law is not the ally of the politician, but his slave, that which it is not the interest of the politician to keep out of the view of the subject is, what the law *is ;*—that which it is his interest to keep (nor even that in all parts) out of the view of the subject, is,—what it is for the interest of the subject that the law *should be ;*—what, in a word, the law ought to be.

Having brought the rule of action within a compass, the narrowness of which, in respect of the quantity of words, has never, regard being had to the amplitude of the matter, yet been equalled,—the tyrant of France has by this one act of charity displayed a quantity of merit, ample enough of itself to form a covering to no inconsiderable a portion of his sins.

But the exemplifications of vague generalities afforded by these systems of classification are sufficiently striking. To save the authors of the systems from ranking any one of the offences in question under a denomination which would be manifestly inapplicable to it, and from the discredit which would attach to them from such a source,—ascending to a superior height in the logical scale,—in the scale of genera and species,—they provide a set of denominations so boundless in their extent, as to be capable without impropriety of including any objects whatsoever on which it might be found convenient to stamp the factitious quality desired.   Noxiousness to other individuals in this or that way, noxiousness to a person himself in this or that way—noxiousness to a particucular class of the community in this or that way—noxiousness to the whole community in this or that way,—these are qualities which it is not in the power of despotism to communicate to any act of any sort; but to cause such persons as it is performed by, to be punished with such or such a punishment, these are effects which, be the sort of act what it may, it is but too easy for supreme power in whatsoever hands reposed, to annex to it.

Here then are so many instances where the turn of the man in power not being capable of being served, or at least so well served, by giving to an object that which is at once its most particular and most proper name, a name of more general and extensive import is employed for the purpose of favouring that deception, which by the designating of it by such its proper name, would have been dissipated, and thus giving to an exercise of power, which, if rightly denominated, would have been seen to be improper and mischievous, the chance of not appearing in such its true light.

CHAPTER IX.

Anti-rational Fallacies.

Ad verecundiam.

Exposition.

WHEN reason is found or supposed to be in opposition to a man's interests, his study will naturally be to render the faculty itself and whatsoever issues from it an object of hatred and contempt.

So long as the Government contains in it any sort of abuse from which the Members of the Government or any of them, derive in any shape a profit, and in the continuance of which, they possess a proportionable interest, reason being against them, persons so circumstanced will be in so far against reason.

Instead of reason we might here say *thought* : Reason is a word that implies not merely the use of the faculty of thinking, but the right use of it. But sooner than fail of its object, the sarcasm and other figures of speech employed upon the occasion are directed not merely against reason, but against *thought* itself: as if there were something in the faculty of thought that rendered the exercise of it incompatible with useful and successful practice.

1. Sometimes a plan the adoption of which would not suit the official person's interest, is without more ado pronounced a *speculative* one : and by this observation all need of rational and deliberate discussion,—

such as objection to the end proposed, as not a fit one,—objection to the means employed as not being fit means,—is considered as being superseded.

To the word *speculative*, for further enforcement, are added or substituted, in a number more or less considerable, other terms, as nearly synonymous to it and to one another, as it is usual forwords called *synonymous* to be : viz. *theoretical, visionary, chimerical, romantic, utopian.*

2. Sometimes a distinction is taken, and thereupon a concession made.   The plan is *good* in *theory*, but it would be *bad in practice, i. e.* its being good in theory does not hinder its being bad in practice.

3. Sometimes, as if in consequence of a further progress made in the art of irrationality, the plan is pronounced to be *too good to be practicable* : and its being so good as it is, is thus represented as the very cause of its being bad in practice.

4. In short, such is the perfection at which this art is at length arrived, that the very circumstance of a plan's being susceptible of the appellation of a *plan,* has been gravely stated as a circumstance sufficient to warrant its being rejected : rejected, if not with hatred, at any rate with a sort of accompaniment, which to the million, is commonly felt still more galling—with contempt.

" Looking at the House of Commons with these views (says a writer on the subject of Parliamentary Reform,) my object would be to find out its *chief* defects, and to attempt the remedy of these *one by one.*

To propose no *system, no great project*, nothing which pretended even to the name of a *plan*, but to introduce in a *temperate* and *conciliatory* manner * * * one or two separate bills [a]."

In this strain were these men proposed to be addressed Anno 1810, by Mr. Brougham : in this strain were they addressed Anno 1819 by Sir James Mackintosh, in moving for a committee on the penal laws. To give a man any chance of doing any thing with them, in this same way they have ever been addressed, and must ever be addressed, till by radical reform (for it cannot be by any thing less) the house shall have been purged of a class of men of whom the most complete inaptitude in respect of every element of appropriate aptitude, is an essential characteristic. In the scale of appropriate probity, in the scale of appropriate intellectual aptitude, to find their level, a man must descend below that of the very dregs of the people. Oh what a picture is here drawn of them, and by so experienced a hand ! How cutting, yet how unquestionably just, the perhaps unintended, perhaps intended satire ! To avoid awakening the real terrors of some, the sham terrors of others, all consistency, all comprehensive acquaintance with the field of action must be abjured. When idolatry in all its shapes shall have become extinct, and the words

---

[a] This was Brougham : the time about June 1810. Reference is made to the Government periodical called the Satirist (by Manners), June 1810, No. 33. p. 570. But that wretched performance is now pretty well forgotten.

wise ancestors no longer an instrument of deception, but a by-word, with what scorn will not ancestors such as these be looked back upon by their posterity?

Intimate as is the connexion between all these contrivances, there is however enough of distinction to render them, in this or that point of view, susceptible of a separate exposure.

Exposure.

Sect. 1. *Abuse of the words* Speculative, Theoretical, &c.

On the occasion of these epithets, and the propositions of which they constitute the leading terms, what will be held up to view in the character of a fallacy, is, not the use of them, but merely the abuse.

It may be placed to the account of *abuse*, as often as in a serious speech, without the allegation of any specific objection, an epithet of this class bestowed upon the measure, is exhibited as containing the expression of a sufficient reason for rejecting it, by putting upon it a mark of reprobation thus contemptuous.

What is altogether out of dispute, is, that many and many a measure has been proposed, to which this class of epithets, or some of them, would be justly applicable. But a man's conceptions must be woefully indistinct, or his vocabulary deplorably scanty, if, be the bad measure what it may, he cannot contrive to give intimation of what, in his view, there is bad in it, without employing an epithet, the effect of which is to hold out, as an object of contempt, the very act of thinking, the operation of *thought* itself.

The fear of theory has, to a certain extent, its foundation in reason. There is a general propensity in those who adopt this or that theory to push it too far: i. e. to set up a general proposition which is not true until certain exceptions have been taken out of it,—to set it up without any of those exceptions,—to pursue it without regard to the exceptions,—and thence, *pro tanto*, in cases in which it is false, fallacious, repugnant to reason and utility.

The propensity thus to push theory too far is acknowledged to be almost universal.

But what is the just inference? Not that theoretical propositions, i. e. propositions of considerable extent, should from such their extent be concluded to be false *in toto* : but only that in the particular case, inquiry should be made, whether, supposing the proposition to be in the character of a general rule generally true, there may not be a case in which, to reduce it within the limits of truth, reason and utility, an exception ought to be taken out of it.

Every man's knowledge is, in its extent, proportioned to the extent as well as number of those general propositions, of the truth of which, they being true, he has the persuasion in his own mind : in other words, the extent of these his theories comprises the extent of his knowledge.

If, indeed, his theories are false, then, in propor-

tion as they are extensive, he is the more deeply
steeped in ignorance and error.

But from the mere circumstance of its being theoretical, by these enemies to knowledge its falsehood is
inferred as if it were a necessary consequence, with
as much reason as if from a man's speaking it were
inferred as a necessary consequence, that what he
speaks must be false.

One would think, that in thinking there were something wicked or else unwise ; every body feels or fan-
cies a necessity of disclaiming it. " I am not given
to speculation." " I am no friend to theories." Speculation, theory, what is it but thinking? Can a man
disclaim speculation, can he disclaim theory, without
disclaiming thought? If they do not mean thought,
they mean nothing ; for unless it be a little more
thought than ordinary, theory, speculation, mean nothing.

To escape from the imputation of meditating destruction to mankind, a man must disclaim every
thing that puts him above the level of a beast.

A plan proposes a wrong end ; or the end being
right, proposes a wrong set of means. If this be what
a man means, can he not say so? Would not what he
says have somewhat more meaning—be a little more
consistent with the principles of common sense—
with common honesty—than saying of it that it is
theoretical—that it is speculative?

Sect. 2. *Utopian.*

As to the epithet *utopian*, the case in which it is rightly applied seems to be that in which, in the event of the adoption of the proposed plan, felicitous effects are represented as about to take place, no causes adequate to the production of such effects being to be found in it.

In Sir Thomas More's romance, from which the epithet utopian has its origin, a felicitous state of things is announced by the very name.

Considering the age in which he lived, even without adverting to the sort of religion of which he was so honest and pertinacious an adherent, we may be sufficiently assured that the institutions spoken of by him as having been productive of this effect, had, taking them altogether, very little tendency to produce it.

Such, in general, is likely enough to be the case with the portion of political felicity exhibited in any other romance : and thus far the epithet *romantic* is likely enough, though not certain, to be found well applied to any political plan, in the conveyance of which, to the notice of the public, any such vehicle is employed. Causes and effects being alike at the command of this species of poet in prose, the honour of any felicitous event is as easily ascribed to *uninfluencing circumstances*, or even to *obstacles*, as to *causes*.

If the established state of things, including the abuse which in so many shapes is interwoven in it, were any thing like what the undiscriminating de-

fenders of it represent it as being, viz. a system of perfection,—in this actually established system, (*real* in
so far as abuse and imperfection are ascribed to it,
imaginary in so far as exemption from such abuse and
imperfection is ascribed to it,) might indeed be seen
an utopia—a felicitous result, flowing from causes not
having it in their nature to be productive of any such
effects, but having it in their nature to be productive
of contrary effects.

In every department of Government, say the advocates of reform, abuses and imperfections are abundant ; because the hands in which the powers of Government are reposed, have, partly by their own artifice, partly by the supineness of the people, been
placed in such circumstances, that abuse in every
shape is a source of profit to themselves.

Under these circumstances, if any expectation were
really entertained, that by these hands any considerable defalcation from the aggregate mass of abuse
will ever be made,—to no other expectation can the
charge of utopianism be with more propriety applied :
effects so produced would be produced against the
force of irresistible obstacles, as well as absolutely
without a cause.

But in that same system there has all along been
preserved, by the many, a faculty, and that a faculty
every now and then, though much too seldom and
too weakly, exercised, of creating, and without very
considerable inconvenience or danger to themselves,
uneasiness, more or less considerable, to these their

rulers. In the state of things thus described, there is
nothing of utopianism; for it is matter of universally
notorious fact; and in this faculty on the part of the
many of creating uneasiness in the bosoms of the few,
—in this faculty on the part of those who suffer by
the abuses of creating uneasiness in the bosoms of
those who profit by them,—in this invaluable, and,
except in America, unexampled faculty,—rests the
only chance, the only source of hope.

Sect. 3. *Good in theory, bad in practice.*

Even in the present stage of civilization, it is almost
a rare case, that by reason, looking to the end in view,
matters of government are determined: and the cause
is, the existence of so many institutions, which being
adverse to the only proper end, the greatest happiness
of the greatest number, are maintained, because favourable to the interests of the ruling few. Custom, blind
custom, established under the dominion of that separate and sinister interest, is the guide by which most
operations have been conducted. In so far as the
interest of the many has appeared to the governing
few to coincide with their own separate interests, in
so far it has been pursued, in so far as it has appeared incompatible with those interests, it has been
neglected or opposed.

One consequence is, that when by accident a plan
comes upon the carpet, in the formation of which the
only legitimate end of Government has been looked

to, if the beaten track of custom has in ever so slight a degree been departed from, the practical man, the man of routine, knows not what to make of it; its goodness, if it be good, its badness, if it be bad, are alike removed out of the sphere of his observance. If it be conducive to the end, it is more than he can see, for the end is what he has not been used to look to.

In the consideration of any plan, what he has not been used to, is to consider what, in the department in question, is the proper end of every plan that can be presented, and whether the particular plan in question be conducive to that end. What he has been used to, is, to consider whether in the matter and form it be like what he has practised. If in a certain degree unlike, it throws him into a sort of perplexity. If the plan be a good one, and in the form of reasons, the points of advantage whereby it is conducive to the proper end in view have been presented, and in such sort that he sees not any, the existence of which he feels himself able to contest, nor at the same time any disadvantages which he can present in the character of preponderant ones, he will be afraid so far to commit himself as to pronounce it a bad one. By way of compounding the matter, and to show his candour, if he be on good terms with you, he will perhaps admit it to be good, viz. in *theory*. But this concession made, it being admitted and undeniable that theory is one thing and practice another, he will take a distinction, and, to pay him for his concession, propose to you to admit that it is not the thing for practice :—in a word, that it is good in theory, bad in practice.

That there have been plans in abundance which have been found bad in practice, and many others which would if tried have proved bad in practice, is altogether out of dispute.

That of each description there have been many which in theory have appeared, and with reference to the judgment of some of the persons by whom they have been considered, have been found *plausible*, is likewise out of dispute.

What is here meant to be denied, is, that a plan which is essentially incapable of proving good in practice can with propriety be said to be good in theory.

Whenever out of a number of circumstances, the concurrence of all of which is necessary to the success of a plan, any one is, in the calculation of the effects expected from it, omitted, any such plan will in proportion to the importance of the omitted circumstance be defective in practice ; and if such be the degree of importance, *bad ;* upon the whole, a bad one ; the disadvantageous effects of the plan not finding a compensation in the advantageous ones.

When the plan for the illumination of the streets by gas-lights was laid before the public by the person who considered himself or gave himself out for the inventor, one of the items in the article of expense—one capital article, viz. that of the pipes, was omitted.

On the supposition that the pipes might all of them
have been had for nothing, and that in the plan so exhibited no other such imperfections were to be found,
the plan would, to the persons engaged in the undertaking, be not merely advantageous, but advantageous
in the prodigious degree therein represented.   If, on
the contrary, the expense of this omitted article were
such as to more than countervail the alleged balance
on the side of profit, then would the plan, with reference to the undertakers, prove disadvantageous upon
the whole, and in one word a bad one.

But whatever it prove to be in practice, in theory,
having so important an omission in it, it cannot but
be pronounced a bad one; for every plan in which, in
the account of advantages and disadvantages, of profit
and losses, any item is on the side of disadvantage or
loss omitted, is, in proportion to the magnitude of
such loss, a bad one, how advantageous soever upon
trial the result may prove upon the whole.

In the line of political economy, most plans that
have been adopted and employed by government for
enriching the community by money given to individuals, have been bad in practice.

But if they have been bad in practice, it is because
they have been bad in theory.   In the account taken
of profit and loss, some circumstance that has been
necessary to render the plan in question advantageous
upon the whole, has been omitted.

This circumstance has been, the advantage which,
from the money employed, would have been reaped,

either in the way of addition to capital by other means, or in the way of comfort by expenditure.

Of the matter of wealth, portions that by these operations were but *transferred* from hand to hand, and commonly with a loss by the way, were erroneously considered as having been *created*.

Sect. 4. *Too good to be practicable.*

There is one case in which, in a certain sense, a plan may be said to be too good to be practicable, and that case a very comprehensive one. It is where, without adequate inducement in the shape of personal interest, the plan requires for its accomplishment that some individual or class of individuals shall have made a sacrifice of his or their personal interest to the interest of the whole. Where it is only on the part of some one individual or very small number of individuals that a sacrifice of this sort is reckoned upon, the success of the plan is not altogether without the sphere of moral possibility; because instances of a disposition of this sort, though extremely rare, are not altogether without example: by religious hopes and fears, by philanthropy, by secret ambition, such miracles have now and then been wrought. But when it is on the part of a body of men or a multitude of individuals taken at random that any such sacrifice is reckoned upon, then it is that in speaking of the plan the term *utopian* may without impropriety be applied.

In this case, if, neglecting the question of practicability, on the mere consideration of the nature of the results the production of which is aimed at by the plan, it can with propriety be termed a good one, the observation, *too good to be practicable*, cannot justly be accused of want of truth.

But it is not any such intimation that, by those in whose mouths this observation is most in use, is meant to be conveyed. The description of persons by whom chiefly, if not exclusively, it is employed, are those who, regarding a plan as being adverse to their interests, and not finding it on the ground of general utility exposed to any preponderant objection, have recourse to this objection in the character of an instrument of contempt, in the view of preventing those from looking into it who might otherwise have been disposed.

It is by the fear of seeing it practised that they are drawn to speak of it as impracticable.

In the character of opposers of a plan, of the good-ness of which (that is, of its conduciveness to the welfare of the whole community taken together) they are themselves persuaded, it cannot be their intention or wish to exhibit themselves: it is not, therefore, in any such property of the plan that it can be their aim to engage those on whom it depends, to look for the cause of the impracticability which they impute to it.

Under favour of such observation as may have been made of the instances in which plans,—the goodness of which, supposing them carried into effect, has been beyond dispute,—have failed of success, what they aim at is the producing, in superficial minds, the idea of a universal and natural connexion between extraordinary and extensive goodness and impracticability : that so often as upon the face of any plan the marks of extraordinary and extensive utility are discernible, these marks may, as it were by a signal, have the effect of inducing a man to turn aside from the plan, and, whether in the way of neglect and non-support, or in the way of active opposition, to bestow on it the same treatment that he would be justified in bestowing upon a bad one.

" Upon the face of it, it carries that air of plausibility that, if you were not upon your guard, might engage you to bestow more or less of your attention upon it. But were you to take the trouble, you would find that, as it is with all these plans that promise so much, practicability would at last be wanting to it. To save yourself from this trouble, the wisest course you can take, is, therefore, to put the plan aside, and think no more about the matter."

There is a particular sort of grin—a grin of malicious triumph—a grin made up of malicious triumph with a dash of concealed foreboding and trepidation at the bottom of it—that forms a natural accompaniment of this fallacy, when vented by any of the sworn defenders of abuse : and Milton, instead of cramming all his angels of the African complexion into the divinity school disputing about predestination, should have employed part of them at least in practising this grin, with the corresponding fallacy, before a looking-glass.

Proportioned to the difficulty of persuading men to regard a plan as otherwise than beneficial, supposing it carried into effect, is the need of all such arguments or phrases as present a chance of persuading them to regard it as impracticable : and according to the sort of man you have to deal with, you accompany it with the grin of triumph, or with the grimace of regret and lamentation.

There is a class of predictions, the tendency and object of which is to contribute to their own accomplishment ; and in the number of them is the prediction involved in this fallacy. When objections on the ground of utility are hopeless or have been made the most of, objections on the ground of practicability still present an additional resource : by these, men who, being convinced of the utility of the plan, are in ever so great a degree well-wishers to it may be turned aside from it : and the best garb to assume for the purpose of the attempt, is that of one who is a well-wisher likewise.

Till the examples are before his eyes, it will not be easy for a man who has not himself made the observation to conceive to what a pitch of audacity political improbity is capable of soaring: how completely, when an opportunity that seems favourable presents itself, the mask will sometimes be taken off :—what thorough confidence there is in the complicity, or in the imbecility of hearers or readers.

If to say *a good thing is a good thing* is nugatory, and as such, foolish language, what shall we say of him who stands up boldly and says, to aim at doing good is a bad thing?

In so many words, it may be questioned whether any such thing has yet been said : but what is absolutely next to it, scarce distinguishable from it, and in substance the same thing, has actually been said over and over. To aim at perfection has been pronounced to be utter folly or wickedness ; and both or either at the extreme. To say that man (the species called *man*) has so much as a tendency to better himself, and that the range of such tendency has no certain limits,—this has been—*speculation :* propositions or observations to that effect have also been set down as a mark of wickedness. " By Priestley an observation to this effect has somewhere or other been made. By Godwin an observation to this effect has somewhere or other been made. By Condorcet, or some other Frenchman or Frenchmen of the class of those who, for the purpose of holding them up to execration, are called philosophers, an observation to this effect has some where or other been made.

" By this mark, with or without the aid of any other, these men, together with other men of the same leaven, have proved themselves the enemies of mankind : and you too, whosoever you are, if you dare to maintain the same heresy, you also are an enemy to mankind."

In vain would you reply to him, if he be an official man, Sir, Mr. Chalmers who, like yourself, was an official man, has maintained this tendency, and written a book which, from beginning to end, is a demonstration of it as clear and undeniable as Euclid's: and Mr. Chalmers is neither a madman nor an enemy to mankind.

In vain would you reply to him, if he call himself a Christian, Sir, Jesus said to his disciples, and to you if you would be one of them, " Be ye perfect, even as our Father in heaven is perfect;" and in so doing has not only assumed the tendency, but commanded it to be encouraged and carried to its utmost possible length.

By observations such as these, may the sort of man in question be perhaps for a moment silenced; but neither by this, nor any thing, nor any body, though one rose from the dead, would he be converted.

To various descriptions of persons over and above those who are in the secret, a fallacy of this class is in a singular degree acceptable and conciliating.

1. To all idle men; all haters of business; a considerable class, where a share in the sovereignty of an empire such as ours is parcelled out into portions which are private property;—where electors' votes are free in appearance only, and scarcely in appearance, and where the votes that are sold for money are in fact among the freest that are to be found.

2. All ignorant men: all who for want of due and appropriate instruction, feeling themselves incapable of judging on any question on its own merits, look out with eagerness for such commodious and reputation-saving grounds.

3. All dull and stupid men: in whose instance, in-

formation—reading—such as has fallen to their lot, has not yet been sufficient to enable them to determine a question on its own merits.

When a train of argument—when but a single argument is presented that requires thought,—an operation so troublesome and laborious as that which goes by the name of *thought*,—an expression of scorn levelled at the author, or supposed author, of this trouble, as far as it goes, a just, howsoever scanty and inadequate, punishment for the disturbance attempted to be given to honourable repose.

Under the name of theory, &c.: what is it that to men of this description is so odious? what but reference to the *end*—to that which on that part of the field of thought and action which is in question, is, or at any rate ought to be, the end pursued—and thence, in every case, the end in view? (how often must it, and ever in vain be repeated?) the greatest happiness of the greatest number. But were reference made to this end—to this inflexible standard —every thing almost they do—every thing almost they support—would stand condemned. What then shall be the standard? Custom—Custom: Custom being their own practice, blindly imitating the practice of men in the same situations, put in motion and governed by the same sinister interests.

CHAPTER X.

Paradoxical Assertion.

Ad judicium.

1. *Dangerousness of the Principle of Utility.*—2. *Uselessness of Classification.*—3. *Mischievousness of Simplification.*—4. *Disinterestedness, a mark of profligacy.*

Exposition.

WHEN of any measure, practice or principle, the utility is too far above dispute to be capable of being impeached by reasoning, a rhetorician to whose interests or views it has appeared adverse, has in some instances in a sort of fit of desperation made this attack upon it; taking up the word or set of words commonly employed for the designation of it, without any such attempt as that of opposing it by any specific objection, he has assailed it with some vehement note of reprobation or strain of invective, in which the mischievousness or folly of it has been taken for granted as if it were undeniable.

Exposure is a sort of process of which the device in question is scarce susceptible : but for the purpose of exposition an example or two may have its use.

Utility, method, simplification, reason, sincerity. By a person unexperienced in the arts of political and verbal warfare, it would not readily be imagined, that entities like these should, by any man laying claim to the distinguishing attribute of man, be pointed out as fit objects of hatred and contempt : yet so it is.

1. As to utility. Already has been named "a great character in a high situation" by whom the principle of utility was pronounced a *dangerous* one[a]. A book

---

[a] Lord Loughborough, when Attorney-General *. The observation made by him in the year 1789, was reported to me presently after it was made. Not till many years after—some time, as I perceive, after what is in the text was written, was the true import perceived by me. The character in which at the time it presented itself to my view, was that of a gross absurdity : its true character was that of deep sagacity. By the principle of utility, what I understood, was—that principle which states as the only justifiable end of Government, the greatest happiness of the greatest number of the members of the community. At that time I still continued to take for granted (such was my simplicity) that this was the end generally aimed at, though often so widely missed   By those, whose desire it was that on each occasion that end should be attained, it could not without self-contradiction be supposed that the endeavour to attain it could seem dangerous. But Loughborough was too well acquainted with the state of Government in this country (not to speak of other countries) not to know that it is the greatest happiness of the ruling few, and not that of the greatest number, that is the end pursued on each occasion by the ruling few. What then was the interest which, on that occasion, as on all occasions, that member of the ruling few, had at heart, and thence preferably if not exclusively, in mind ? It was, of course, the interest of the ruling few. But the interest of the ruling few is, on the greatest part of the field of Government, in a state of continued opposition to that of the greatest number : accordingly a principle, which, in case of competition, and to the extent of the competition, called for sacrifices to be made of the interests of the class to which he belonged, and which alone was the object of his solicitude, could not but, in his eyes, be a dangerous one.

---

* Note written August 21, 1819 :—"Great characters in high situations," a phrase employed about December 1809 by minister Percival, then Attorney-General, in calling down the vengeance of the law on I forget what alleged libeller, for using the appellation of "the Doctor"

might be mentioned, and one of no small celebrity[a], in which the same principle—the principle of utility —has been pronounced useless:—the principle itself, and consequently every investigation in which to the purposes of legislation or common life, application is endeavoured to be made of it.

What must be acknowledged is, that to make a right and effectual use of it, requires the concurrence of those requisites which are not always found in company: invention, discernment, patience, sincerity; each in no inconsiderable degree; while, for the pronouncing of decisions without consulting it, decisions in the *ipse dixit* style, nothing is required but boldness.

Not that, on any occasion on which it promises to suit his purpose, and he feels in himself a capacity to apply it to that purpose, the most decided scorner of it, ever fails to make use of it. It is only when, if consulted, its decisions would be against him, or he feels himself awkward at consulting it, that he ever takes upon him to do without it: and to prove any thing to be right or wrong, thinks it sufficient for himself to say so.

2.—*Classification a bad thing.*—*Good Method a bad thing.*

On the same occasion in which a convenience was found in pronouncing the principle of utility useless,

---

in speaking of Lord Sidmouth. By the picture it seemed to give of the character of the prosecution and of the persons who bore a part in it, it made a lasting impression on my mind.

[a] The Edinburgh Review.

the like convenience was found in professing the like contempt for that quality in discourse which goes by the name of good method, or, simply, *method*, and that sort of operation called good classification, or, simply, *classification*.

When the subject a man undertakes to write upon, is to a certain degree extensive, as, for example, the science of morals, or that of legislation, whether what a man says be clear or not, of falsehood, will depend upon the goodness of the method in which the parts of it have been cast.   1. If for example, snow and charcoal were both classed under the same name and neither of them had any other, if the question were asked, whether the thing known by that name were white or black, no inconsiderable difficulty would be found in answering it either by a yes or no.   2. And if under favour of the identity of denomination, sugar of lead were to be used in a pudding instead of any of the sort of sugar usually applied to that purpose, practical inconveniences analogous to those which were experienced by Thornbury from eating pancake[a], might probably be found to result from the mistake thus exemplified in the tactical branch of the art or science of life, call it which you please.

In the course of an attempt made[b] to cast the

---

[a] In the pancake in question, which, at a table at the Cape of Good Hope, was served up to a company of which Thornbury, better known by his travels in Japan, was one, white lead was employed instead of flour :—some recovered, and some died.

[b] In a book written Anno 1781, published Anno 1789, under the title of Introduction to Morals and Legislation.

whole multitude of pernicious actions into apt classes, —as a fruit, and proof, and test of the supposed aptitude, about a dozen propositions were mentioned as being capable of being without any deviation from the truth of things, ascribed to the pernicious acts, respectively, collected together under one denomination by the names respectively assigned to the four classes to which they were referred.

On the same occasion intimation was likewise given, that in the system of law and law terms in use, for the designation of offences, among English lawyers, no such fair general denomination could be found, to the contents of which an equal number, and it might perhaps have been added, any number at all of common propositions could, without error and falsehood, be ascribed.   A system of classification and nomenclature which can never be employed without confounding at every turn, objects which, to prevent practical and painful accidents, require to be distinguished, must by every man, who has not a decided interest in maintaining the contrary, be acknowledged to be very ill adapted to those which are, or at least which ought to be, its purposes.

Here then was an intimation given, that the whole system of English penal law is in an extreme degree ill adapted to what ought to be the purposes of every system of law : and an implied invitation to those, if any such there were, who being conversant in the subject of law, had any desire to see it well adapted to its professed purposes, to show that the system was not in respect of the points indicated, a bad one, the radically bad one it was there represented to be, or else to take measures for making it better. But it being the interest of every one who is most conversant with this subject, that the whole system, instead of being as good as it can be made, should be as bad as those who live under it will endure to see it, the invitation could not in either branch be accepted.

In any other branch of science that can be named, Medicine, Chemistry, Natural History in all its branches, the progress made in every other respect is acknowledged to be commensurate to, and at once effect and cause in relation to the progress made in the art of classification:—nor in any one of those branches of science, would it, perhaps, be easy to find a single individual by whom the operation of classification would be spoken of as any thing below the highest rank in the order of importance. Why this difference? Because in any one of these branches of science there is scarce an individual to whose interest the advancement of the science is opposed : whereas among the professors of the law there exists not an individual to whose interest the advancement of the art of legislation is not opposed ;—is not, either immediately detrimental or ultimately dangerous.

3. *Simplification.*

By the opposite vice, *complication*, every evil opposite to the ends of justice, viz., uncertainty of the law itself, unnecessary delay, expense and vexation in respect of the execution, is either produced or aggra-

vated[a].   Consequently, to every one by whom any
wish is entertained of seeing the mass of these evils
reduced, a fervent desire is entertained of seeing the
virtue of simplification infused into the system of law
and judicial procedure.   On an occasion that took
place not long ago, if the account of the debates can
be trusted, a gentleman was found resolute and frank
enough to stand up and rank this virtue, if after that,
such it may be called, among the worst of vices : the
use of it was evidence of Jacobinism : evidence of the
circumstantial kind indeed, but sufficiently conclusive.

If on a declaration to that effect any sentiment of
disapprobation were visible in the language or deportment of that honourable house, none such are, at
least, recorded : and if none such really were perceptible, this circumstance alone might afford no inconsiderable ground for the desire expressed by some, of
seeing the character of that honourable house under-
go a thorough change.

4. *Disinterestedness a mark of profligacy.*

In his pamphlet on his Official Economy Bill, to
give up official emolument, is by Edmund Burke pronounced, in so many words, to be "a mark of the
basest profligacy :"

On somewhat more defensible grounds might this
position itself be pronounced as strong a mark as ever

---

[a] See Scotch Reform Table.

was exhibited, or ever could be exhibited, of the most shameless profligacy.

An assumption contained in it, besides others too numerous to admit of their being detailed here, is— that in the eyes of man there is nothing that has any value—nothing that is capable of actuating and giving direction to his conduct, but the matter of wealth: that the love of reputation and the love of power are themselves, both of them, without efficient power over the human heart.

So opposite is this position of his to the truth, that the less the quantity of money which, in return for his engagement to render official service, a man, not palpably unfit for the business of it, is content to accept, the stronger is the proof, the presumptive evidence thereby afforded, of his aptitude in all points, with relation to the business of that office : since it is a proof of his relish for the business—of the pleasure he anticipates from the performance of it[a].

Blinded by his rage, in this his frantic exclamation, wrung from him by the unquenched thirst for lucre, this madman, than whom none perhaps was ever more mischievous,—this incendiary, who contributed so much more than any other to light up the flames of that war, under the miseries occasioned by which the nation is still groaning, poured forth the reproach of "the basest profligacy" on the heads of thousands, before whom,

---

[a] See Bentham par Dumont. *Traité des Peines et des Recompenses:* and Defences of economy against Edmund Burke and George Rose.

had he known who they were, he would have been
ready to bow the knee.   Not to mention the whole
magistracy of the empire, whose office is that of justice of the peace,—among other persons before whom
he was in the habit of prostrating himself, of the verbal filth he thus casts around him, one large mass
falls upon the head of the Marquess Camden, and
from his, rebounds upon those other official heads,
from which the surrender made of the vast mass of
official emolument, drew forth the stream of eulogium
which the documents of the day present us with.

5. *How to turn this fallacy to account.*

To let off a paradox of this sort with any chance of
success, you must not be any thing less than the leader
of a party.   For if you are, instead of gaping and
staring at you, men will but laugh at you, or think of
something else without so much as laughing at you,
because there is no laughing at any thing without
thinking of it.

Moreover a thing of this sort succeeds much better
in a speech, than in a book or pamphlet; and that, for
several reasons.

The use of a speech is to carry the measure of the
moment : and if the measure be but carried, no matter for the means.   The measure being carried, the
paradox is seen to be no less absurd and mischievous
than it is strange : no matter ; the measure is carried.
War is declared, or a negotiation for peace broken
off.   Peace you will have some time or other, but in

the mean time the paradox has had its effect. A law has passed : and that law an absurd and mischievous one. Some day or other the mischief may receive a remedy : but that day may not arrive these two or three hundred years [a].

In a speech too, it is all profit, no loss. Your point may be gained, or not gained : your reputation remains where it was. It is your speech, or not your speech, whichever is most convenient. To A, who, under the notion of its being yours, admires it, it *is* your speech; to B, who, because it is yours, or because it is an absurd and mischievous one, spurns at it, it is *not* your speech. If the words of your paradox are ambiguous, as they will be, if they are well and happily chosen,— susceptible of two senses, an innoxious and a noxious one ; this is exactly what is wanted. *A*, who on your credit is ready to take it, and to adopt it in the noxious one which suits your purpose, is suffered silently to take it in that noxious one. But if *B*, taking it in the noxious one, attacks you and pushes you too hard, then some adherent of yours (not you yourself, for it would be weak indeed for you to appear in the matter), some adherent of yours brings out the innocent sense, vows and swears it was *that* meaning that was yours, and belabours poor B with a charge of calumny.

If in the choice of your expression you have been

---

[a] Till lately, the country has suffered in a variety of ways by the law made in the reign of Elizabeth to prevent good workmanship : the effect is felt, the cause men cannot bear to look at.

negligent or unfortunate, so that no more than one sense, and that one indefensible, can with any colour of reason be ascribed to it, you thus lose part of your advantage: but still no harm can happen to you; you disavow, that is your adherent, for you, disavows the very words : and thus every thing is as it should be.

Thus it is that from speeches—spoken and unminuted speeches—you derive much the same sort of advantage as is derived from that sort of sham law (which, in so far as it is made by any body, is made by judges, and is called common or unwritten law) by lawyers : thundering all the while the charge of insincerity or folly in all who have the assurance to ascribe to it either a different word, or a different meaning. To the supposed speech, as to the supposed law, they give what words they please, and then to those words they give what meaning they please. The law, indeed, neither has, nor ever had, any determinate form of words belonging to it: whereas the speech could not have been spoken, unless it had had a set, and that a complete one, of determinate words belonging to it. But in the speech—the words never having been committed to writing, or if they have been, evidence of their being the same words not being producible,—the speech-maker is as safe as if he had never uttered any one of those words.

In the intellectual weakness of those on whom, in this form, imposition is endeavoured to be practised —in this degrading weakness, and in the state of servitude in which they are accordingly held by the shackles of authority[a], may be seen the cause of that success, and thence of the effrontery and insolence, which this species of imposition manifests. In proportion as intellect is weaker and weaker, reason has less and less hold upon it; authority, fortified by the appearance, real or fallacious, of strong persuasion, more and more.

It is in this way that, strange as at first mention it cannot but appear—it is in this way—and when addressed to minds of such a texture, the more flagrant and outrageous the absurdity, the stronger its persuasive force. Why? because without the strongest ground, a persuasion—so strong a persuasion, of the truth of a proposition, at first sight at least, so adverse to truth, it is taken for granted, could not have been formed.

When the terrors of which religion is the source, are the instruments employed for inculcating it, the strength of the persuasion thus inspired, presents little cause for wonder. In the intensity of the exertion made for the purpose of believing, the greater the difficulty, the greater is, in case of success, the merit. Hence that most magnanimous of all conclusions, *credo quia impossibile est.* Higher than this, the force of faith—the force and consequently the merit—cannot go: by this one bound the pinnacle is attained; and whatsoever reward Omnipotence has in store for

---

[a] See 1. *Ad Verecundiam,* Ch. 1. Fallacy of Authority.

service of this complexion, is placed out of the reach of failure.

Be the absurdity ever so flagrant,—the nature of man considered, and how absolute the dominion which is exercised over him by the passions of fear and hope, — be the absurdity ever so flagrant, cause of just wonder can never be afforded, by any acceptance which it receives with the support afforded to it by the most irresistible of the passions :

The understanding is not the source, reason is of itself no spring of action : the understanding is but an instrument in the hand of the will : it is by hopes and fears that the *end* of action is determined ; all that reason does, is to find and determine upon the *means :*

But where, at the mere suggestion of a set of men with gowns of a certain form on their backs, where at their mere suggestion (unsupported by any motive of a nature to act on the will), we see men living and acting under the persuasion that in the vice of lying, there is virtue to metamorphose into justice, the crime of usurpation ;—here it is not the will that is confounded and overwhelmed, it is the understanding that is deluded[a].

---

[a] To form a ground for decision, a judge asserts as true, some fact which to his knowledge is not true : some fact for the assertion of which, if, in the station of a witness, and without having for his protection the power of a judge, a man were to venture the assertion of, he would by this same judge be punished with imprisonment and infamy. To screen it from the abhorrence due to it, this lie, exceeding in wickedness the most wicked of the assertions commonly brought into view under that name, is decked up in the same appellation, *fic-*

tion, which is employed in bringing to view the innoxious and amusing pictures of ideal scenes for which we are indebted to the poetic genius. What you are thus doing with the lie in your mouth?—had you power to do it without the lie?—Your lie is a foolish one. Have you no such power?—It is a flagitious one. In this mire may be seen laid the principal part of the foundation of English common law.

CHAPTER XI.

*Non-causa pro causâ : or, Cause and obstacle con-
founded.*

Ad judicium.

Exposition.

WHEN in a system which has good points in it you
have a set of abuses, or any of them to defend ; after
a general eulogium bestowed on the system, or an indication more or less explicit of the good effects the
existence of which is out of dispute, take the abuses
you have to defend, either separately or collectively,
(collectively is the safest course,) and to them ascribe
the credit of having given birth to the good effects.

Cùm hoc, ergò propter hoc.

In every political system which is of long standing,
and which not having been produced, any considerable
part of it, in prosecution of any comprehensive design, good or bad, but piece-meal at different and distant times, according to the casual and temporary
predominance of conflicting interests, whatsoever may
be the good or the bad points in the state of things
which at any given time constitutes the result of it,
among the incidents which may be observed as having
place in it, some, upon proper scrutiny and proper
distinction made, may be seen to have operated in
the character of effective or promotive causes; others,

in the character of obstacles or preventives; others, to have been in relation to them, in the character of immaterial incidents or inoperative circumstances.

In such a system, whatsoever are the abuses or other imperfections in it, and whatsoever are the prosperous results observable in it, these prosperous results will have found, in the abuses and imperfections, not so many efficient or promotive causes, but so many obstacles or preventives.

Meantime, if so you can order matters that, instead of being recognised as having operated in the character of obstacles, the abuses in question shall be believed to have operated in the character of efficient or promotive causes, nothing can contribute more powerfully to the effect which it is your endeavour to produce.

If you cannot so far succeed as to cause the prosperous results in question to be referred to the abuses by which they have been obstructed and retarded, the next thing you are to endeavour at, is, to cause them to be ascribed to some inoperative circumstance or circumstances, having in appearance some connexion or other—the nearer the better—with the abuses.

At any rate, you will, as far as depends upon you, cause the prosperous circumstances in question to be referred to any causes rather than the real ones: for in proportion as it becomes manifest of what causes they are the results, it will become manifest of what other circumstances they have not been the results: whereupon, no sooner is any one of the abuses you have to defend, considered in this point of view, than a question will be apt to occur :—Well, and this ?— what has been the use of this ? To which no answer being found, the consequence is such as need not be mentioned.

Real knowledge being among the number of your most formidable adversaries, your endeavour must of course be to obstruct its advancement and propagation as effectually as possible.

Real knowledge depends in a great degree on the being able, on each occasion, to distinguish from each other, causes, obstacles, and uninfluencing circumstances : these, therefore, it must on every occasion be your study to confound as effectually as possible.

Exposure.

Example 1.—*Good Government :—Obstacle represented as a cause,—the influence of the Crown.*

If the superiority of the Constitution of the English limited monarchy, as compared with all absolute or less limited monarchies, be in England a point undisputed, and regarded as indisputable, and the characteristic by which that limited monarchy is distinguished from all absolute and less limited monarchies, is, the influence, the superior influence of the mass of the people,—the influence exercised by the will of the nominees of the people on the wills of the nominees of the king, and thence on the conduct of the king himself,—a circumstance which, in so far as it operates, diminishes the efficiency of this influence, and on many, if not most occasions, may be seen to destroy that efficiency altogether,—cannot with propriety be numbered among the causes of that superiority, but must, on the contrary, be placed to the account of the obstacles that obstructed it.

In point of fact, the members of the House of Commons, some really, all in supposition, nominees of the mass of the people, act, as to the nominees of the king,—viz. the members of the executive department,—with the authority of judges : viz. to the purpose of causing *punishment* to be inflicted under the name of punishment, in case of special delinquency, not without the concurrence of the House of Lords ; but to the purpose of causing *removal*, without any such concurrence.

In so far as over the will of the nominees of the people as above mentioned, acting in their above-mentioned character of judges, an efficient influence is exercised by the king or his nominees, the efficiency of this judicial authority is destroyed ; the nominees of the king, in the exercise of their respective functions, committing any enormities at pleasure; and thereupon in the character though without the name of judges, absolving themselves, and, if such be their pleasure, praising themselves for what they have done.

In this case, the fallacy consists in representing, defending, and supporting, in the character of an indispensable cause of the acknowledged prosperous results, the sinister and corruptive influence in question;

—a circumstance which, so far from being in any degree a promotive cause, is an obstacle.

In what way it operates in the character of an obstructive and destructive circumstance, has already been shown above. In what way, with relation to the same effect, it can operate as a cause, has never been so much as attempted to be shown : it has been on every occasion taken for granted, and this, on no other ground than that of its being a concomitant circumstance.

Example 2.—*Effect, good Government :—Obstacle represented as a cause,—station of the bishops in the House of Lords.*

To good government, neither in the situation of a bishop nor in any other situation can a man be contributory, any further than as he takes a part in it.

In that department of government which is carried on in the House of Lords, a man cannot bear a part any further than as he takes a part in the debates carried on there, or at least attends and gives his vote.

But of the whole body of bishops, including since the Union those from Ireland, a small part, upon an average scarce so many as a tenth, are seen to attend and give their votes : and as for speaking—when any instance of it happens to take place, it sets men a-staring and talking as if it were a phenomenon.

How comes it that the number of those who vote,

and especially of those who speak, is so small?   Because a general feeling exists, that to that class temporal occupations and politics are not suitable occupations.

And why not suitable?

1. Because in that war of personalities in which, in a large proportion, the debates in that as well as in the other House consist, a man of this class is in a peculiar degree vulnerable.   The Apostles, did they bear any part in, had they any seat in, the Roman Senate, or so much as in the Common-council of the city of Jerusalem?   Was it Peter, was it James, was it John, —was it not Dives, that used to clothe himself in purple and fine linen?—Walking from place to place to preach, comprised their occupations.   If yours were the same, would you not be rather more like them than you are?

2. Because there is a general feeling, though not expressed in words, from a sort of decency and compassion, that a legislative assembly is not a fit place for a man who is not at liberty to speak what he thinks; and who, should he be bold enough to bring to view any one of the plainest dictates of political utility, might be put to silence and confounded by reference to this or that one of the 39 Articles, or by this or that text of Scripture, out of a Testament Old or New.

So many things of which, however improbable, he is bound to profess his belief.

So many things which, however indefensible by

reason, he would be bound, were he to open his mouth, to defend.

Matter of duty to him to be,—matter of infamy not to be,—steeled against conviction.

So many vulnerable parts with which he is embarrassed, and with which an antagonist of his is not embarrassed.

So many chains with which he is shackled, and with which an antagonist of his is not shackled.

A man whose misfortune should it be to hear a word or two of reason, it would be his duty not to listen to it.

To a man thus circumstanced, to talk reason would have something ungenerous in it and indecorous : it would be as if a man should set about talking indecently to his daughter or his wife.

In vain would they answer, what has been so often answered, that neither Jesus nor his Apostles ever meant what they said—that every thing is to be explained and explained away. By answers of this sort, those and those alone would be satisfied, whose satisfaction, with every thing that is established, is immoveable, and not susceptible of experiencing diminution from any objections, or increase from any answers.

Example 3.—*Effect, useful national learning ; Obstacle stated as a cause, system of education pursued in Church-of-England Universities.*

On the subject of learning, to the question whether with relation to it the universities might with more propriety be considered as causes or as obstacles, much need not here be said, after what has been said on the subject by the Reverend Vicesimus Knox, and of late by the Edinburgh Review.

If these fragments, with the exception of the scurrilous parts of the Review, were put together and made into a book, a most instructive addition to it might be made by a history of the treatment experienced from this quarter by the inventions of the quaker Lancaster. In the age of academical and right-reverend orthodoxy, learning, it would there be seen, is even to the very first rudiments of it, an object of terror and hatred.

Of this Quaker, though he undertook not to attempt to make converts, what is certain, is, that no school would, under his management, have been a school of perjury : and since, in so far as by his means the elementary parts of knowledge made their way among the people, intellectual light would take place of intellectual darkness, he experienced the hostility that might so naturally have been expected from those who love darkness better than light, to wit, for a reason which may be seen in that book, the knowledge of which it was his object to diffuse, as it was theirs to confine and stifle it.

In virtue and knowledge, in every feature of felicity, the empire of Montezuma outshines, as every body knows, all the surrounding states, even the Commonwealth of Tlascala not excepted.

Where (said an inquirer once, to the high priest of

the temple of Vitzlipultzli), where is it that we are to look for the true cause of so glorious a pre-eminence? "Look for it!" (answered the holy pontiff,) "Where shouldst thou look for it, blind sceptic, but in the copiousness of the streams in which the sweet and precious blood of innocents flows daily down the altars of the great God?"

"Yes," answered in full convocation and full chorus the archbishops, bishops, deans, canons, and prebends of the religion of Vitzlipultzli:—"Yes," answered in semi-chorus the vice-chancellor, with all the doctors, both the proctors and masters regent and non-regent of the as yet uncatholiciz'd university of Mexico:—"Yes, in the copiousness of the streams in which the sweet and precious blood of innocents flows daily down the altars of the great God."

Example 4.—*Effect, national virtue; Obstacle represented as a cause, opulence of the clergy.*

In several former works it has been shown [a], that, be the effect what it may,—in so far as money or in any other shape the matter of reward, is, in the character of an efficient cause, employed in the view or under the notion of promoting it,—what degree of efficiency shall attend in such case the use made of the instrument, depends not so much upon its magnitude

---

[a] *Traité des Peines et des Recompenses.* Defence of Economy against Burke, and Do. against Rose: both in the Pamphleteer, Anno 1817. Church-of-Englandism Examined, 1818.

as upon the manner in which, and the skill with which, it is applied; and in particular, that in so far as that instrument is composed of public money, it is no less possible, and in some cases much more frequent, so to apply it that the production of that effect shall, instead of being promoted, be prevented : that when, as for working, a man is paid alike whether he does work or whether he does none, to expect work from him is impossible, and to pretend to expect it, mere mockery : that after engaging to render an habitual course of service (for the rendering of which no extraordinary degree of talent or alacrity is necessary), a fit person has received that which is necessary to obtain his free engagement for the rendering it, every penny added has no other tendency than to afford him means and incentives to relinquish his duties for whatever other occupations are more suitable to his taste.

Now if this be true of all men; it is true of every man : and it is not a man's being called prebend, canon, dean, bishop, or even archbishop, that will in his case or in any other person's case make it false.

It is a proposition that, be it ever so true, is not evident, but requires argument deduced from experience to render it so, that by such service as is rendered by the English clergy, virtue is in any degree promoted.

It is a proposition that, be it to a certain extent ever so true, is to a certain extent notoriously not true, that to the procurement of such service, money from any source is necessary. For without a particle of money passing from hand to hand, service of this sort is rendered by men one towards another, viz. among the people called Quakers : and if for the exhibiting to view the comparative degrees of efficiency with which service of this sort is rendered,—work of this sort done,—who is there that will take upon him to deny that the highest degree of the scale would be found occupied by the people called Quakers, or disputed with them by the people called Methodists, while the very lowest would be recognised as being occupied without dispute by the members sacred or profane of the established and most opulently endowed Church of England.

It is another proposition that still remains to be proved, that, admitting that for the procurement of this service—to the whole extent in which for the production of virtue it is wanted,—money is necessary, it is also necessary that for the raising of the necessary quantity, money should by the power of Government be forced out of the pockets of unwilling contributors.

CHAPTER XII.

Partiality-preacher's Argument.

Ad judicium.

From the abuse, argue not against the use.

Exposition.

FROM abuse it is an error (it has been said) to argue against use.

The proposition is an absurd one, make the best of it, but the degree of absurdity will depend upon the turn that may be given to the sentence.

Whichsoever be the turn given to it, the plain and undeniable truth of the case as between use and abuse will alike serve for the exposure of it.

Be the institution what it may, whatsoever good effects there are that have resulted from it, these constitute, as far as experience goes, the *use* of it : whatsoever ill effects have resulted from it, these, in so far at least as they have been the object of foresight and the result of intention, constitute the *abuse* of it.

Thus as to past results : and the same observation applies to expected future ones.

Exposure.

Now then come the fallacies to the propagation of which it may and must have been directed.

1. In taking an account of the effects of an institu-

tion, you ought to set down all the good effects and omit all the bad ones.

This is one of the purposes to which it is capable of being applied: this needs not much to be said of it.

2. In taking an account of the effects of an institution good and bad, you ought not to argue against it on the supposition that the sum of the bad ones is greater than the sum of the good ones, merely from the circumstance that among all its effects taken together, there are some that belong to the bad side of the account.

In this latter sense, such is the character of the maxim that nothing can be said against the truth of it. As an instruction, it is too obvious to be of any use: in the way of warning, it cannot by possibility do any harm, nor is it altogether out of the sphere of possibility, that in this or that instance it may have its use.

Applied to a man's pecuniary affairs it amounts to this: viz. Conclude not that a man has no property because he has some debts.

CHAPTER XIII.

The End justifies the Means.

Ad judicium.

IN this case surely, if in any, exposition is of itself exposure.

The insertion of this article in the list of fallacies was suggested by the use made of it in the Courier Newspaper of the 27th of August 1819, as reported and commented upon in the Morning Chronicle of the 28th[a].

The end justifies the means. Yes: but on three conditions, any of which failing, no such justification has place.

1. One is, that the end be good.

2. That the means chosen be either purely good, or if evil, having less evil in them than on a balance there is of real good in the end.

---

[a] The Courier Newspaper is, in the other public prints, perpetually spoken of as enjoying the favour of the Monarch of the day. I have all along been upon the watch to see whether a denial in any shape of that assertion would be given. I have never been able to hear of any such thing. The fact admitted, a conclusion which can scarcely be refused is, that the principles manifested in that paper are the principles entertained and acted upon by that royal arbiter of our fate, in whose disposal the lives and fortunes of about twenty millions or thereabout in the three kingdoms, and sixty millions in Asia, are placed. Without deigning to wait for and receive, or if received, to have regard to the evidence on the other side, at the solicitation of Lord Sidmouth, Secretary of State, the Prince

3. That they have more of good in them, or less of evil, as the case may be, than any others, by the employment of which the end might have been attained.

Laying out of the case these restrictions, note the absurdities that would follow.

Acquisition of a penny loaf is the end I am at. The goodness of it is indisputable. If by the goodness of the end, any means employed in the attainment of it are justified, instead of a penny, I may give a pound for it: thus stands the justification on the ground of prudence. Or, instead of giving a penny for it, I may cut the baker's throat, and thus get it for nothing: and thus stands the justification on the ground of benevolence and beneficence.

In politics, what is the use of this fallacy? In the mouth of one whose station is among the INS, it will serve for whatsoever cruelties those by whom power is exercised may at any time find a pleasure in committing on those over whom power is exercised, for

Regent by one letter dated August 1819 bestows his approbation upon the conduct maintained by the Manchester Magistrates, on the occasion of the slaughter committed by their officers—by the armed yeomanry on an unarmed multitude: and by another, dated the same month, upon Sir John Bing, the General Commander of the Regulars, for the support given by him to it. What shall we say of this? Let prudence give the answer. The Secretary is worthy to serve such a Sovereign: the Sovereign is worthy to be served by such a Secretary.—Every stroke he adds to his own portrait, the faithful servant adds to that of his royal patron and protector. A complete portrait thus formed by lines copied from the Courier, would constitute a most instructive and interesting piece.

the purpose of confirming themselves in the power of committing more such cruelties.

The INS, as such, have the power to commit atrocities, and that power having sinister interest for its spur, is never suffered to be idle. For the use of this fallacy, in so far as it can be worth their while to employ a cloak, they have therefore a continual demand.

The OUTS, acting under the impulse of the same spur, sharpened by continual privation, and continually repeated disappointment, have on their part a still more urgent demand for the same fallacy, though the opportunities of making application of it but rarely present themselves to their hands.

The oracular party adage—invented by the Whigs : —Not men but measures, or Not measures but men : —for according as you complete the sentence, you may word it either way,—This bold but slippery instrument of fallacy has manifest alliance with the present. Seating in office fit men, being the end, every thing depending upon that end, and the men in question being the only ones by which it can be attained, no means can be imagined by which such an end may not be justified.

CHAPTER XIV.

Opposer-General's justification :—Not measures but men; or, Not men but measures.

Ad invidiam.

ACCORDING to the notions commonly entertained of moral duty under the head of probity, and in particular under the head of that branch of probity which consists in sincerity, whatsoever be the nature and extent of the business in question, private or public, it is not right for a man to argue against his own opinion ;—when his opinion is so and so, to profess it to be the reverse, and in so doing to bend the force of his mind to the purpose of causing others to embrace the opinion thus opposite to his real one.

That, in particular, if being a member of the House of Commons, and in opposition, a measure which to him seems a proper one, is brought on the carpet on the ministerial side, it is not right that he should declare it to be, in his opinion, pernicious, and use his endeavours to have it thought so, and treated as such by the House; and so again, if, being on that same side, a measure such as to him appears pernicious, is brought on the carpet on the side of opposition, it is not right that he should declare it to be, in his opinion, beneficial and fit to be adopted, and accordingly use his endeavours to make it generally thought so, and as such adopted by the House.

An aphorism, said to have been a favourite one with the late Mr. Charles Fox, is the proposition at the head of this chapter.

Not men but measures! or, Not measures but men! are the two forms in either of which, according as the ellipsis is filled up, the aphorism may be couched.

Not measures but men! is the more simple expression of the two, it being in that form that the aphorism is marked out for approbation : reprobation being the sentiment attached to its opposite.   *Not men but measures!*

If you look to speeches, then comes the constant and constantly interminable question—what were the words in the speeches. The words are in that case on each occasion genuine or spurious, the interpretation correct or incorrect, according as it suits the purpose of him who is speaking of it, and more particularly of him who spoke it, that it should be.

But on one occasion we have the aphorism from the pen of Charles Fox himself: and then, if applied to the question of sincerity or insincerity, as above, it is found to have no direct bearing on it.

" *Are to be attended to,*" are the words employed on this occasion to complete the proposition. " How vain, how idle, how presumptuous (says the declaimer in his attempt to put on the historian) is the opinion that laws can do every thing ! and how weak and pernicious the maxim founded upon it, that measures not men, are to be attended to !"

Weak enough as thus expressed, it must be confessed: and abundantly too weak to be by a statesman considered as worth noticing even by so vague and ungrounded a note of reprobation.—As if any one ever thought of denying that both ought to be "*attended to!*" and as if, even in a debating club, words so vague and unmeaning as "*attended to*" were a fit subject of debate.

What must be confessed is, that to a man who wishes well to his country, and sees a set of men who in his opinion are a bad set, conducting the affairs of it, few things are more provoking than by this or that comparatively unimportant, but so far as it goes beneficial measure, to see them obtain a degree of reputation of which one effect may be to confirm them in their seat.

But what seems not to have been sufficiently "*attended to*" is, that it is by the badness of their measures that the only warrant for giving to the men the appellation of bad men can be grounded : that if they are really the bad men they are supposed to be, have a little patience, and they will come out with some bad measure, against which, it being by the supposition bad, and by yourself looked upon as such, you may without prejudice to your sincerity, point your attacks : and if no such bad measure ever came from them, the imputation of their being bad men, is rather premature.

Distressing indeed to a man of real probity must be the alternative : to see a set of men fixed in this their all-commanding seat, and making a proportionally extensive and pernicious use of it; or, for the purpose of taking what chance is to be had of precluding them from this advantage, to keep on straining every endeavour to make the House and the public look upon as pernicious, a measure of the utility of which he is himself satisfied.

In the abomination of long and regularly corrupt parliaments lies the cause of this distress.

Under this system, when the whole system of abuses has a determined patron on the throne, and that patron has got a set of ministers that suit this ruling purpose, misrule may swell to such a pitch, that without any one measure in such sort bad that you can fix upon it and say this is a sufficient ground for punishment, or even for dismission, the State may be at the brink of ruin :—meantime some measure may be introduced, against which, though good or at least innoxious of itself, the people, by means of some misrepresentation of matter of fact, or some erroneous opinion or other which prevails among them, may to the disgrace and expulsion of the ministry be turned against it, and then comes the distressing alternative.

But were the duration of the assembly short, and the great and surely effective mass of the matter of corruption expelled and kept out of it, no such alternative would ever present itself. The chance of ridding the country of a bad set of ministers would be renewed continually. The question supposed to be tried on each occasion might be the question really tried : whereas at present on each occasion the question tried is but one and the same, viz. Shall the ministry or shall it not continue?

The question brought on the carpet is like the wager in a feigned issue, a mere farce, which, but for its connection with the principal question above mentioned, would not be deemed worth trying, and would not be tried.

CHAPTER XV.

Rejection instead of Amendment.

Ad judicium.

Exposition.

THIS fallacy consists in urging in the character of a bar, or conclusive objection against the proposed measure, some consideration, which, if presented in the character of a proposed amendment, might have more or less claim to notice.

It generally consists of some real or imaginary inconvenience, alleged commonly, but not necessarily, as eventually to result from the adoption of the measure.

This inconvenience, supposing it real, will either be preponderant over the promised benefit or not preponderant.

In either case it will be either remediable or irremediable.

If at the same time irremediable and preponderant, then it is, and then only, that in the character of an objection it is of itself conclusive.

By him in whose mind discernment and candour are combined, this distinction will be not only felt, but brought to view. If in respect of adequate discernment there be a failure, it will not be felt: if in respect of candour only, it will have been felt, but it will not be brought to view.

The occasion by which opportunity is afforded for

the working of this fallacy, is the creation of any new office, including the mass of emolument which, without inquiry into the necessity, or any means taken for keeping down the quantum of it within the narrowed limits which the good of the service admits of, is, by the union of habit with the sinister interest that gave birth to it, annexed as of course, upon their creation, to all new offices.

The fallacy, what there is of fallacy in the case consists in the practice of setting up the two universally applicable objections, viz. *need of economy*, and mischief or danger from the increase of the *influence of the crown*, in the character of peremptory bars to the proposed measure.

Exposure.

The ground on which an objection of this stamp may with propriety be considered and spoken of under the denomination of a fallacy, is where the utility of the proposed new establishment is left unimpeached, and the sole reason for the rejection proposed to be put upon the proposed measure consists in the above topics or one of them.

In such case, on the part of him by whom any objections so inconclusive in their nature are relied on, the reliance placed on them amounts to a virtual acknowledgement of the utility of the proposed new establishment: inasmuch as in an address from one rational being to another, nothing seems, upon the face of the statement at least, more unnatural, than that if

a man could find any objection that would apply to
the particular establishment in question in contradistinction to all others, he should confine himself to
an objection which applies alike to almost all existing
establishments ; that is, to almost the whole frame of
the existing government.

Such is the case where the two common-place objections in question, or either of them, are brought
out in the character of objections by themselves, and
without being accompanied by any specific ones.

But even when added to specific ones, an objection
thus inconclusive in its nature, if urged in a direct
way, and dwelt upon with any emphasis, can scarcely,
at least while there remain any useless places unabo-
lished, or any overpaid places, from which the overplus of emolument remains undefalcated, be exempted
from the imputation of irrelevancy.

At any rate, wherever it happens that a minister at
present in office sees opposite to him in the House
another person who has at any time been in office, it
seems an observation not very easy to answer in the
character of an argument *ad hominem,* should it be
said, " When you were in office, there were such and
such offices which were of no manner of use ; these
you never used your endeavours to abolish, notwithstanding the use that would have resulted from the
abolition, in the shape of diminution of needless expenditure and sinister influence : yet now, when a set
of offices is proposed, for which you cannot deny but
that there is *some* use, your exertions for the benefit

of economy are reserved to be directed against these useful ones."

No doubt but that on the supposition that the two opposite masses of advantage and disadvantage being completely in equilibrio,—advantage in the shape of service expected to be rendered in the proposed new offices on the one hand, disadvantage in the shape of expense of the emolument proposed to be attached to them on the other,—a weight much less than that of the mischief from the increase of sinister influence, would suffice to turn the scale.

Take also another supposition. Suppose (what is not in every case possible) that the value of the service expected to be obtained by means of the proposed new offices is capable of being obtained, and has accordingly been obtained in figures. Suppose on the other hand (what will very frequently be feasible) that the expense of the establishment may with sufficient precision be obtained in figures, and being so obtained, on striking the balance, found to be less than the advantage so expected from the service. Suppose lastly, (what is impossible) that the value of the mischief which, in the shape of introduction of additional influence, were with sufficient precision capable of standing expressed in figures had been so expressed, and being so expressed, the quantity of mischief in this shape were found sufficient to turn the scale on the side of disadvantage.

Here would be a sufficient reason for the rejection of the proposed establishment, and thence a sufficient warrant for bringing into the field the argument in question, common-place as it is. But in regard to this last supposition at any rate, how far it is from being capable of being realized, is but too evident.

Upon the whole, therefore, so far at least as concerns the objection drawn from the increase that would result to the sinister influence of the crown, it may be said that whatsoever time is spent in descanting upon this topic may be set down to the account of lost time.

It is a topic, the importance of which is surely sufficient to entitle it to be considered by itself. The influence of the crown, it ought always to be remembered, can no otherwise receive with propriety the epithet sinister, than in so far as, by being directed to and reaching a member of Parliament or a parliamentary elector, it affects the purity of Parliament. But by a system of measures properly directed to that end, the constitution of Parliament might be effectually guarded against any degree of impurity capable of being productive of any sensible inconvenience, whatsoever were the lucrativeness of the utmost number of offices, for the creation or preservation of which so much as a plausible reason could be found: and were it otherwise, the proper remedy would be found, not in the refusal to create any new office, the service of which was understood to overbalance in any determinate and unquestionable degree the mischief of the expense, but in the taking the nomination out of the hands of the crown, and vesting it in some other and independent hands.

2 A

The putting all places in these respects upon the same footing—necessary and unnecessary ones,—properly paid and overpaid ones,—wears out and weakens that energy which should be reserved for, and directed with all its force against, unnecessary places, and the overplus part of the pay of overpaid ones.

Another occasion on which this fallacy is often wont to be applied, is the case in which, from the mere observation of a profit as likely from any transaction to accrue to this or that individual, a censure is grounded, pronouncing it a *job*.

The error in case of sincerity, the fallacy in case of insincerity, consists, in forgetting that individuals are the stuff of which the public is made ; that there is no way of benefiting the public but by benefiting individuals ; and that a benefit which, in the shape of pleasure or exemption from pain, does not sooner or later come home to the bosom of at least some one individual, is not in reality a benefit—is not entitled to that name.

So far then from constituting an argument in disfavour of the proposed measure, every benefit that can be pointed out as accruing or likely to accrue to any determinate individual or individuals, constitutes, as far as it goes, an argument in favour of the measure.

In no case whatsoever—on no imaginable supposition—can this consideration serve with propriety in the character of an argument in disfavour of any measure. In no case whatsoever—on no imaginable supposition—can it, so far as it goes, fail of serving with propriety in the character of an argument in favour of the measure. Is the measure good ?—It adds to the mass of its advantages.   Is the measure upon the whole a bad one ?—It subtracts, by the whole amount of it, from the real amount of the disadvantages attached to the measure.

At the same time in practice, there is no argument, perhaps, which is more frequently employed, or on which more stress is laid, without doors at any rate, if not within doors, than this, in the character of an argument in disfavour of a proposed measure : no argument which, even when taken by itself, is with more confidence relied on in the character of a conclusive one.

To what cause is so general a perversion of the faculty of reason to be ascribed ?

Two causes present themselves as acting in this character :

1. It is apt to be received (and that certainly not without reason) in the character of evidence—conclusive evidence—of the nature of the motive, to the influence of which the part taken by the supporters of the measure, or some of them, (viz. all who in any way are partakers of the private benefit in question,) ought to be ascribed.

In this character, to the justness of the conclusion thus drawn, there can in general be nothing to object.

But the consideration of the motive in which the part taken either by the supporters or the opposers of a measure finds its cause, has elsewhere been shown to be a consideration altogether irrelevant[a]; and the use of the argument has been shown to be of the number of those fallacies, the influence of which is in its natural and general tendency unfavourable to every good cause.

The other cause is the prevalence of the passion of envy. To the man to whom it is an object of envy, the good of another man is evil to himself. By the envy of the speaker or writer, the supposed advantage to the third person is denounced in the character of an evil, to the envy of the hearers or the readers:— denounced, and perhaps without any perception of the mistake, so rare is the habit of self-examination, and so gross and so perpetual the errors into which, for want of it, the human mind is capable of being led.

In speaking of the passion or affection of *envy*, as being productive of this fallacious argument, and of the error, but for which shame would frequently restrain a man from the employment of it, it is not meant to speak of this passion or this affection as one of which, on the occasion in question, the influence ought to be considered as pernicious on the whole.

So far from being pernicious, the more thoroughly it is considered, the more closely it will be seen to be salutary upon the whole; and not merely salutary, at least in the best state of things that has yet been realized, but so necessary, that without it, society would hardly have been kept together.

---

[a] See Part 2. Personalities.

The legislator who resolves not to accept assistance from any but social motives, from none, save what in his vocabulary pass under the denomination of pure motives, will find his laws without vigour and without use.

The judge who resolves to have no prosecutors who are brought to him by any but pure motives, will not find that part of his emolument which, under the present system of abuse, is composed of fees, and may save himself the trouble of going into court—of sitting on penal causes. The judge who should determine to receive no evidence but what was at the same time brought to him, and, when before him, guided by pure motives, need scarcely trouble himself to hear evidence.

The practical inference is—that, if he would avoid drawing down disgrace upon himself instead of upon the measure he is opposing, a man ought to abstain from employing this argument in confutation of the fallacy; since, in as far as he employs it, he is employing in refutation of one fallacy (and that so gross an one, that the bare mention of it in that character may naturally be sufficient to reduce the employer to silence), he is employing another fallacy, which is of itself susceptible of a refutation no less easy and conclusive.

It is only by the interests, the affections, the passions (all these words mean nothing more than the same psychological object appearing in different characters), that the legislator, labouring for the good and in the service of mankind, can effect his purposes. Those interests, acting in the character of motives, may be of the self-regarding class, the dissocial, or the social :—the social he will, on every occasion where he finds them already in action, endeavour not only to engage in his service, but cherish and cultivate : the self-regarding and the dissocial, though his study will be rather to restrain than encourage them, he will at any rate, wherever he sees them in action or likely to come into action, use his best endeavours to avail himself of directing their influence, with whatever force he can muster, to his own social purposes.

PART THE FIFTH.

CHAPTER I.

Characters common to all these Fallacies.

Upon the whole, the following are the characters which appertain in common to all the several arguments here distinguished by the name of fallacies :

1. Whatsoever be the measure in hand, they are, with relation to it, irrelevant.

2. They are all of them such, that the application of these irrelevant arguments, affords a presumption either of the weakness or total absence of relevant arguments on the side on which they are employed.

3. To any good purpose they are all of them unnecessary.

4. They are all of them not only capable of being applied, but actually in the habit of being applied, and with advantage, to bad purposes : viz. to the obstruction and defeat of all such measures as have for their object and their tendency, the removal of the abuses or other imperfections still discernible in the frame and practice of the government.

5. By means of their irrelevancy, they all of them consume and misapply time, thereby obstructing the course, and retarding the progress of all necessary and useful business.

6. By that irritative quality which, in virtue of their irrelevancy, with the improbity or weakness of which it is indicative, they possess, all of them, in a degree more or less considerable, but, in a more particular degree such of them as consist in personalities, they are productive of ill-humour, which in some instances has been productive of bloodshed, and is continually productive as above, of waste of time and hindrance of business.

7. On the part of those who, whether in spoken or written discourses, give utterance to them, they are indicative either of improbity or intellectual weakness, or of a contempt for the understandings of those on whose minds they are destined to operate.

8. On the part of those on whom they operate, they are indicative of intellectual weakness: and on the part of those in and by whom they are pretended to operate, they are indicative of improbity, viz. in the shape of insincerity.

The practical conclusion is, that in proportion as the acceptance and thence the utterance of them can be prevented, the understanding of the public will be strengthened, the morals of the public will be purified, and the practice of government improved.

CHAPTER II.

Of the mischief producible by Fallacies.

THE first division that presents itself in relation to the mischief of a fallacy, may be expressed by the words specific and general.

The specific mischief of a fallacy, consists in the tendency which it has to prevent or obstruct the introduction of this or that useful measure in particular.

The general mischief, consists in that moral or intellectual depravation which produces habits of false reasoning and insincerity :—this mischief may again be distinguished into mischief produced *within doors* and mischief produced *without doors*.

Under the appellation of mischief within doors, is to be understood all that mischief, that deception, which has its seat in the bosom of any member of the supreme legislative body.

Under the appellation of mischief without doors, all that which has its seat in the bosom of any person not included in that body,—of any person whose station is among the people at large.

CHAPTER III.

Causes of the utterance of these Fallacies.

THE causes of the utterance of these fallacies may, it should seem, be thus denominated and enumerated.

1. Sinister interest—self-conscious sinister interest.
2. Interest-begotten-prejudice.
3. Authority-begotten-prejudice.
4. Self-defence, *i. e.* sense of the need of self-defence against counter fallacies.

First Cause.

Sinister interest, of the operation of which the party affected by it is conscious.

The mind of every public man is subject at all times to the operation of two distinct interests ; a public and a private one. His public interest is that which is constituted of the share he has in the happiness and well-being of the whole community, or of the major part of it: his private interest is constituted of, or by, the share he has in the well-being of some portion of the community less than the major part: of which private interest the smallest possible portion is that which is composed of his own individual—his own personal—interest.

In the greater number of instances, these two interests are not only distinct, but opposite : and that to such a degree, that if either be exclusively pursued, the other must be sacrificed to it.

Take for example pecuniary interest: It is the personal interest of every public man, at whose disposal public money extracted by taxes from the whole community is placed, that as large a share as possible, and if possible the whole of it, should remain there for his own use: it is at the same time the interest of the public, including his own portion of the public interest, that as small a share as possible, and if possible no part at all, remain in these same hands for his personal or any other private use.

Taking the whole of life together, there exists not, nor ever can exist, that human being in whose instance any public interest he can have had, will not in so far as depends upon himself, have been sacrificed to his own personal interest. Towards the advancement of the public interest all that the most public-spirited, which is as much as to say the most virtuous of men can do, is to do what depends upon himself towards bringing the public interest, that is his own personal share in the public interest, to a state as nearly approaching to coincidence, and on as few occasions amounting to a state of repugnance, as possible with his private interests.

Were there ever so much reason for regretting it, the sort of relation which is thus seen to have place between public and private interest, would not be the less true: nor would it be the less incumbent on the legislator, nor would the legislator, in so far as he finds it reconcileable to his personal interest to pursue the public interest, be the less disposed and determined to act and shape his measures accordingly.

But the more correct and complete a man's conception of the subject is, the more clearly will he understand, that in this natural and general predominance of personal, over every more extensive interest, there is no just cause for regret. Why? Because upon this predominance depends the existence of the species, and the existence of every individual belonging to it. Suppose for a moment the opposite state of things—a state in which every one should prefer the public to himself—and the consequences—the necessary consequences, would be no less ridiculous in idea, than disastrous and destructive in reality.

In the ordinary course and strain of legislation, no supposition inconsistent with this only true and rational one, is acted upon. On this supposition is built whatsoever is done in the application made either of the matter of reward, or of the matter of punishment, to the purposes of government. The supposition is—that on the part of every individual whose conduct it is thus endeavoured to shape and regulate, interest, and that, private interest, will be the cause by the operation of which his conduct will be determined: not only so, but that in case of competition as between such public and such private interest, it is the private interest that will predominate.

If the contrary supposition were acted upon, what would be the consequence? that neither in the shape of reward, nor in the shape of eventual punishment, would the precious matter of good and evil be wasted or exposed to waste, but (in lieu of requisition, with reward or punishment, or both, for its sanction, for securing compliance) advice and recommendation would be employed throughout the system of law penal as well as remuneratory.

Thence it is that, in so far as in the instance of any class of men, the state of the law is such as to make it the interest of men belonging to that body to give rise or continuance to any system of abuse however flagrant, a prediction that may be made with full assurance is, that the conduct of that body, that is, of its several members with few or no exceptions, will be such as to give rise or continuance to that system of abuse : and if there be any means which have been found to be, or promise to be conducive to any such end, such means will, accordingly, how inconsistent soever with probity in any shape, and in particular in the shape of sincerity, be employed.

A common bond of connection, says Cicero somewhere, has place among all the virtues : To the word *virtue*, substitute the word *abuse*, meaning abuse in government, and the observation will be no less true. Among abuses in government, besides the logical *commune vinculum* composed of the common denomination *abuse*, there exists a moral *commune vinculum* composed of the particular and sinister interest in which all men who are members of a government so circumstanced have a share.

So long then as any man has any, the smallest particle of this sinister interest belonging to him—so long has he an interest, and consequently a fellow-feeling with every other man who in the same situation has an interest of the like kind. Attack one of them, you attack all ; and in proportion as each of them feels his share in this common concern dear to him, and finds himself in a condition to defend it, he is prepared to defend every other confederate's share with no less alacrity than if it were his own. But it is one of the characteristics of abuse, that it can only be defended by fallacy. It is, therefore, the interest of all the confederates of abuse to give the most extensive currency to fallacies, not only to such as may be serviceable to each individual, but also to such as may be generally useful. It is of the utmost importance to them to keep the human mind in such a state of imbecility, as shall render it incapable of distinguishing truth from error.

Abuses, that is to say institutions beneficial to the few, at the expense of the many, cannot openly, directly, and in their own character, be defended. If at all, it must be in company with, and under the cover of other institutions to which this character either does not in fact appertain, or is not seen to appertain.

For the few who are in possession of power, the principle the best adapted, if it were capable of being set to work, would be that which should be applicable to the purpose of giving to the stock of abuses established at each given period, an unlimited increase.

No longer than about a century ago a principle of this cast actually was in force, and that to an extent that threatened the whole frame of society with ruin : viz. under the name of the principle of *passive obedience* and *non-resistance*.

This principle was a *primum mobile*, by the due application of which, abuses in all shapes might be manufactured for use to an amount absolutely unlimited.

But this principle has now nearly, if not altogether, lost its force. The creation of abuses has, therefore, of necessity been given up ; the preservation of them is all that remains feasible : it is to this work that all exertions in favour of abuse have for a considerable time past, and must henceforward be confined.

Institutions, some good, some bad—some favourable to both the few and the many ; some favourable to the few alone, and at the expense of the many— are the ingredients of which the existing system is composed. He who protects all together, and without discrimination, protects the bad. To this object the exertions of industry are still capable of being directed with a prospect of success : and to this object they actually do continue to be directed, and with a degree of success disgraceful to the probity of the few by whom such breach of trust is practised, and to the intellect of the many by whom it is endured.

If the fundamental principle of all good government, viz. that which states as being on every occasion the proper, and the only proper end in view and object of pursuit, the greatest happiness of the greatest

number, were on every occasion set up as the mark;
on each occasion the particular question would be,
by what particular means can this general object be
pursued with the greatest probability of success?

But by the habit of recurring to and making application of this one principle, the eye of the inquirer,
the tongue of the speaker, and the pen of the writer,
would, on every part of the field of legislation, be
brought to some conclusion, passing condemnation on
some or other of those abuses, the continuance of
which has this common interest for its support.

In a word, so long as any one of these relatively
profitable abuses continues unremedied,—so long
must there be one such person or more to whose interest the use of reason is prejudicial, and to whom
not only the particular beneficial measure from which
that particular abuse would receive its correction, but
every other beneficial measure, in so far as it is
supported by reason, will also be prejudicial in the
same way.

It is under the past and still existing state of
things,—in other words, under the dominion of usage,
custom, precedent, acting without any such recurrence
to this only true principle,—that the abuses in question have sprung up. Custom, therefore, blind custom, in contradiction and opposition to reason, is the
standard which he will on every occasion endeavour to set up as the only proper, safe, and definable
standard of reference. Whatever is, is right: every
thing is at it should be. These are his favourite

maxims—maxims which he will let slip no opportunity of inculcating to the best advantage possible.

Having, besides his share in the sinister interest, his share in the universal and legitimate interest, there must, to a corresponding extent, be laws and institutions, which, although good and beneficial, are no less beneficial to and necessary to his interest, than to that of the whole community of which he makes a part. Of these, then, in so far as they are necessary to his interest, he will be as sincere and strenuous a defender, as of those by which any part of the abuses which are subservient to his sinister interest is maintained.

It is conducive, for instance, to his interest, that the country should be effectually defended against the assault of the common enemy: that the persons and properties of the members of the community in general, his own included, should be as effectually as possible protected against the assaults of internal enemies—of common malefactors.

But it is under the dominion of custom, blind or at best purblind custom—that such protection has been provided. Custom, therefore, being sufficient for his purpose, Reason always adverse to it, Custom is the ground on which it will be his endeavour to place every institution, the good as well as the bad. Referred to general utility as their standard, shown to be conformable to it by the application of reason to the case, they would be established and supported,

2 B

indeed, on firmer ground than at present. But by placing them on the ground of utility, by the application of reason, he has nothing to gain, while, as hath been seen above, he has every thing to lose and fear from it.

The principle of general utility he will accordingly be disposed to represent in the character of "*a dangerous principle*:" for so long as blind custom continues to serve his purpose, such, with reference to him and his sinister interest, the principle of general utility really is.

Against the recognition of the principle of general utility, and the habit of employing reason as an instrument for the application of it, the leading members of the Government, in so far as corruption has pervaded the frame of Government, and in particular the members of all ranks of the profession of the law, have the same interest as in the eyes of Protestants and other non-catholics, the Pope and his subordinates had at the time and on the occasion of the change known in England by the name of the *Reformation*.

At the time of the Reformation, the opposition to general utility and human reason was conducted by fire and sword. At present, the war against these powers can not be completely carried on by the same engines.

Fallacies, therefore, applied principally to the purpose of devoting to contempt and hatred those who apply the principle of general utility on this ground, remain the only instruments in universal use and re-quest for defending the strongholds of abuse against hostile powers.

These engines we, accordingly, see applied to this purpose in prodigious variety, and with more or less artifice and reserve.

CHAPTER IV.

Second Cause.

Interest-begotten Prejudice.

IF by interest in some shape or other, that is by a motive of one sort or other, every act of the will and thence every act of the hand is produced, so directly or indirectly must every act of the intellectual faculty : though in this case the influence of the interest, of this or that motive, is neither so perceptible nor in itself so direct as in the other.

But how (it may be asked) is it possible that the motive a man is actuated by can be secret to himself? Nothing is more easy—nothing more frequent. Indeed the rare case is, not that of his not knowing, but that of his knowing it.

It is with the anatomy of the human mind, as with the anatomy and physiology of the human body: the rare case is, not, that of a man's being unconversant, but that of his being conversant with it.

The physiology of the body is not without its difficulties: but in comparison of those by which the knowledge of the physiology of the mind has been obstructed, the difficulties are slight indeed.

Not unfrequently, as between two persons living together in a state of intimacy, either or each may possess a more correct and complete view of the motives by which the mind of the other, than of those by which his own mind is governed.

Many a woman has in this way had a more correct and complete acquaintance with the internal causes by which the conduct of her husband has been determined, than he has had himself.

The cause of this is easily pointed out. By interest, a man is continually prompted to make himself as correctly and completely acquainted as possible with the springs of action, by which the minds of those are determined on whom he is more or less dependent for the comfort of his life.

But by interest he is at the same time diverted from any close examination into the springs by which his own conduct is determined.

From such knowledge he has not, in any ordinary shape, any thing to gain,—he finds not in it any source of enjoyment.

In any such knowledge he would be more likely to find mortification than satisfaction. The purely social motives, the semi-social motives, and, in the case of the dissocial motives, such of them as have their source in an impulse given by the purely social or by the semi-social motives [a]; these are the motives, the prevalence of which he finds mentioned as matter of praise in the instance of other men : it is by the supposed prevalence of these amiable motives that he finds reputation raised, and that respect and good-will

---

[a] See Introduction to Morals and Legislation.

in which every man is obliged to look for so large a
portion of the comfort of his life.

In these same amiable and desirable endowments he
finds the minds of other men actually abounding and
overflowing; abounding during their lifetime by the
testimony of their friends, and after their departure by
the recorded testimony enregistered in some monthly
magazine, with the acclamation of their friends, and
with scarce a dissenting voice from among their enemies.

But the more closely he looks into the mechanism
of his own mind, the less of the mass of effects produced he finds referable to any of those amiable and
delightful causes; he finds nothing, therefore, to attract
him towards this study, he finds much to repel him
from it.

Praise and self-satisfaction, on the score of moral
worth, being accordingly hopeless, it is in intellectual
that he will seek for it. " All men who are actuated
by regard for any thing but self, are fools; those only,
whose regard is confined to self, are wise. I am of
the number of the wise."

Perhaps he is a man with whom a large proportion
of the self-regarding motives may be mixed up with
a slight tincture of the social motives operating upon
the private scale. What in this case will he do? In
investigating the source of a given action, he will in
the first instance set it down, the whole of it, to the
account of the amiable and conciliatory, in a word,
the social motives. This, in the study of his own

mental physiology, will always be his first step, and this will commonly be his last. Why should he look further? Why take in hand the probe? Why undeceive himself, and substitute a whole truth, that would mortify him, for a half truth that flatters him?

The greater the share which the motives of the social class have in the production of the general tenour of a man's conduct, the less irksome it seems evident this sort of psychological self-anatomy will be. The first view is pleasing; and the more virtuous the man, the more pleasing is that study, which to every man has been pronounced the proper one.

But the less irksome any pursuit is, the greater, if the state of faculties, intellectual and active permit, will be a man's progress in it.

CHAPTER V.

Third Cause.

Authority-begotten Prejudice.

Prejudice is the name given to an opinion of any sort, on any subject, when considered as having been embraced without sufficient examination : it is a judgment, which being pronounced *before* evidence, is, therefore, pronounced without evidence.

Now, at the hazard of being deceived, and by deception led into a line of conduct prejudicial either to himself or to some one to whom it would rather be his wish to do service, what is it that could lead a man to embrace an opinion without sufficient examination ?

One cause is the uneasiness attendant on the labour of examination : he takes the opinion up as true, to save the labour that might be necessary to enable him to discern the falsity of it.

Of the propensity to take not only facts but opinions upon trust, the universality is matter of universal observation. Pernicious as it is in some of its applications, it has its root in necessity, in the weakness of the human mind. In the instance of each individual, the quantity of opinion which it is possible for him to give acceptance or rejection to, on the ground of examination performed by himself, bears but a small proportion to that in which such judgment, as he passes upon it, cannot have any firmer or other ground than that which is composed of the like judgment pronounced by some other individual or aggregate of individuals: the cases, in which it is possible for his opinion to be *home-made,* bear but a small proportion to the cases in which, if any opinion at all be entertained by him, that opinion must necessarily have been imported.

But in the case of the public man, this necessity forms no justification either for the utterance or for the acceptance of such arguments of base alloy, as those which are represented under the name of fallacies.

These fallacies are not less the offspring of sinister interest because the force of authority is more or less concerned. Where authority has a share in the production of them, there are two distinguishable ways in which sinister interest may also have its share.

A fallacy which, in the mouth of A., had its root immediately in interest,—in self-conscious sinister interest,—receiving utterance from his pen or his lips, obtains, upon the credit of his authority, credence among acceptors in any multitude. Having thus rooted itself in the minds of men, it becomes constitutive of a mass of authority, under favour of which, such fallacies as appear conducive to the planting or rooting in the minds of men in general the erroneous notion in question, obtain, at the hands of other men, utterance and acceptance.

2. Having received the prejudice at the hands of authority, viz. of the opinion of those whose adherence to it was produced immediately or mediately by the operation of sinister interest,—sinister interest operating on the mind of the utterer or acceptor of the fallacy in question, prompts him to bestow on it, in the character of a rational argument, a degree of attention exceeding that which could otherwise have been bestowed on it; he fixes, accordingly, his attention on all considerations, the tendency of which is to procure for it utterance or acceptance, and keeps at a distance all considerations by which the contrary tendency is threatened.

CHAPTER VI.

Fourth Cause.

Self-defence against counter Fallacies.

THE opposers of a pernicious measure may be sometimes driven to employ fallacies, from their supposed utility as an answer to counter fallacies.

" Such is the nature of men (they may say), that these arguments, weak and inconclusive as they are, are those which on the bulk of the people (upon whom ultimately every thing depends) make the strongest and most effectual impression : the measure is a most mischievous one : it were a crime on our parts to leave unemployed any means not criminal that promise to be contributory to its defeat. It is the weakness of the public mind, not the weakness of our cause, that compels us to employ such engines in the defence of it."

This defence might, indeed, be satisfactory where the fallacies in question are employed,—not as *substitutes* but only as *supplements* to relevant and direct ones.

But if employed as supplements, to prove their being employed in that character and in that character only, and that the use thus made of them is not inconsistent with sincerity, two conditions seem requisite.

1. That arguments of the direct and relevant kind

be placed in the front of the battle, declared to be the
main arguments, the arguments and considerations by
which the opposition or support to the proposed measure was produced ;

2. That on the occasion of employing the fallacies
in question, an acknowledgement should be made of
their true character, of their intrinsic weakness, and
of the considerations which, as above, seemed to impose on the individual in question the obligation of
employing them, and of the regret with which the
consciousness of such an obligation was accompanied.

If, even when employed in opposition to a measure
really pernicious, these warnings are omitted to be
annexed to them, the omission affords but too strong
a presumption of general insincerity. On the occasion
in question, a man would have nothing to fear from
any avowal made of their true character. Yet he
omits to make this avowal. Why ?—Because he fore-
sees that, on some other occasion or occasions, arguments of this class will constitute his sole reliance.

The more closely the above considerations are ad-
verted to, the stronger is the proof which the use of
such arguments, without such warnings, will be seen
to afford of improbity or imbecility, or a mixture of
the two, on the part of him by whom they are employed : of imbecility of mind, if the weakness of such
arguments has really failed of becoming visible to him :
of improbity, if, conscious of their weakness, and of
their tendency to debilitate and pervert the faculties,
intellectual and moral, of such persons as are swayed

by them,—he gives currency to them unaccompanied
by such warning.

Is it of the one or of the other species of imperfection, or of a mixture of both, that such deceptious
argumentation is evidentiary ? On this occasion, as on
others, the answer is not easy, nor fortunately is
it material, to estimate the connexion between these
two divisions of the mental frame ; so constantly and
so materially does each of them exert an influence on
the other, that it is difficult for either to suffer but the
other must suffer more or less along with it. On many
a well-meaning man this base and spurious metal has
no doubt passed for sterling ; but if you see it bur-
nished, and held up in triumph by the hands of a man
of strong as well as brilliant talents—by a very master
of the mint—set him down, without fear of injuring
him, upon the list of those who deceive without having
any such excuse to plead as that of having been deceived.

CHAPTER VII.

Use of these Fallacies to the utterers and acceptors of
them.

BEING all of them to such a degree replete with
absurdity—many of them upon the face of them composed of nothing else—a question that naturally presents itself is, how it has happened that they have acquired so extensive a currency?—how it is that so
much use has been made, and continues to be made,
of them.

Is it credible (it may be asked) that, to those by
whom they are employed, the inanity and absurdity
of them should not be fully manifest?—Is it credible,
that on such grounds political measures should proceed?

No, it is not credible : to the very person by whom
the fallacy is presented in the character of a reason—
of a reason on the consideration of which his opinion
has been formed, and on the strength of which his
conduct is grounded,—it has presented itself in its
genuine colours.

But in all assemblies in which shares in power are
exercised by votes, there are two descriptions of persons whose convenience requires to be consulted,—
that of the speakers and that of the hearers.

To the convenience of persons in both these situations, the class of arguments here in question are in

an eminent degree favourable : 1. As to the situation
of the speaker :—the more numerous and efficient the
titles to respect which his argument enables him to
produce, the more convenient and agreeable is that
situation made to him. Probity in the shape of independence—superiority in the article of wisdom—superiority in the scale of rank—of all these qualities,
the reputation is matter of convenience to a man ; and
of all these qualities, the reputation is by these arguments promised to be made secure.

. 1. As to independence :—when a man stands up to
speak for the purpose of reconciling men to the vote
he purposes to give, or for the purpose of giving, to
the side which he espouses whatsoever weight is regarded by him as attached to his authority ; the nature
of the purpose imposes on him a sort of necessity of
finding something in the shape of a reason to accompany and recommend it.

Though in fact directed and governed by some
other will behind the curtain, and by the interest by
which that other will is governed, decency is understood to require, that it is from his own understanding,
not from the will of any other person, that his own
will should be understood to have received its direction.

But it is not by the matter of *punishment* or the
matter of *reward*—it is not by *fears* or *hopes*—it is
not by *threats* or *promises*—it is by something of the
nature, or in the shape at least of a reason, that *understanding* is governed and determined. To show,

then, that it is by the determination of his own judgment that his conduct is determined, it is deemed advisable to produce some observation or other in the
character of the determinate reason, from which on the
occasion in question, his judgment, and thence his will
and active faculty, have received their direction.

The argument is accordingly produced, and by this
exhibit, the independent character of his mind is established by irrefragable evidence.

To this purpose every article in the preceding cata-
logue may with more or less effect be made to serve,
according to the nature of the case.

2. Next, as to superiority in the scale of *wisdom :*
—on running over the list, different articles will be
seen to present in this respect different degrees of convenience.

Some of them will be seen scarcely putting in any
special title to this praise.

In others, while the reputation of prudence is secured, yet it is that sort of prudence which, by the
timidity attached to it, is rendered somewhat the less
acceptable to an erect and commanding mind.

To this class may be referred the arguments *ad
metum* and *ad verecundiam*,—the hydrophobia of innovation, the argument of the ghost-seer, whose nervous
system is kept in a state of constant agitation by the
phantom of Jacobinism dancing before his eyes,—the
idolator, who, beholding in ancestry, in authority, in
allegorical personages of various sorts and sizes, in
precedents of all sorts, in great characters dead and

living, placed in high situations, so many tyrants to
whose will, real or supposed, blind obsequiousness at
the hands of the vulgar of all classes, may by apt ceremonies and gesticulation be secured, makes himself
the first prostration, in the hope and confidence of
finding it followed by much and still more devout
prostration, on the part of the crew of inferior idola-
tors, in whose breasts the required obsequiousness has
been implanted by long practice.

Other arguments again there are, in and for the
delivery of which the wisdom of the orator places
itself upon higher ground. His acuteness has pene-
trated to the very bottom of the subject—his comprehension has embraced the whole mass of it—his adroit-
ness has stripped the obnoxious proposal of the delusive colouring by which it had recommended itself to
the eye of ignorance : he pronounces it speculative,
theoretical, romantic, visionary : it may be good in
theory, but it would be bad in practice : it is too good
to be practicable, the goodness which glitters on the
outside is sufficient proof, is evidence, and that conclusive, of the worthlessness that is within : its apparent facility suffices to prove it to be impracticable.
The confidence of the tone in which the decision is
conveyed, is at once the fruit and the sufficient evidence of the complete command which the glance of
the moment sufficed to give him of the subject in all
its bearings and dependences. By the experience
which his situation has led him to acquire, and the
use which his judgment has enabled him to make of

that experience, he catches up at a single glance those features which suffice to indicate the class to which the obnoxious proposal belongs.

By the same decision delivered in the same tone, superiority of rank is not less strikingly displayed than superiority of talent. It is no new observation how much the persuasion, or at least the expression given to it, is strengthened by the altitude of the rank as constituted or accompanied by the fullness of the purse.

The labour of the brain, no less than that of the hand, is a species of drudgery which the man of elevated station sees the propriety and facility of turning over to the base-born crowd below—to the set of plodders whom he condescends upon occasion to honour with his conversation and his countenance. By his rank and opulence he is enabled in this, as in other ways, to pick and choose what is most congenial to his taste. By the royal hand of Frederic, philosophers and oranges were subjected to the same treatment and put to the same use. The sweets, the elaboration of which had been the work of years, were elicited in a few moments by the pressure of an expert hand.

The praise of the receiver of wisdom is always inferior to that of the utterer; but neither is the receiver, so he but make due profit of what he receives, without his praise.

The advantage he acquires from these arguments, is, that of being enabled to give the reason of the faith that is, or is supposed to be, in him.

In some circumstances in which silence will not serve a man, it will, and to a certainty, be construed into a confession of self-convicting consciousness;—consciousness that what he does is wrong and indefensible;—that what he gives men to understand to be his opinion, is not really his opinion;—that of the supposed facts, which he has been asserting to form an apparent foundation for his supposed opinion, the existence is not true.

By a persuasion to any such effect, on the part of those with whom he has to do, his credit, his reputation, would be effectually destroyed.

Something, therefore, must be said, of which it may be supposed that, how little soever may be the weight properly belonging to it, it may have operated on his mind in the character of a reason. By this means his reputation for wisdom is all that is exposed to suffer;—his reputation for probity is saved.

Thus, in the case of this sort of base argument, as sometimes in the case of bad money, each man passes it off upon his neighbour, not as being unconscious of its worthlessness—not so much as expecting his neighbour to be really insensible of its worthlessness—but in the hope and expectation that the neighbour, though not insensible of its worthlessness, may yet not find himself altogether debarred from the supposition, that to the utterer of the base argument, the badness of it may possibly not have been clearly understood.

But the more generally current in the character of an argument any such absurd notion is, the greater is the apparent probability of its being really entertained: for there is no notion, actual or imaginable, that a man cannot be brought to entertain, if he be but satisfied of its being generally or extensively entertained by others.

CHAPTER VIII.

Particular demand for Fallacies under the English Constitution.

Two considerations will suffice to render it apparent that, under the British Constitution, there cannot but exist, on the one hand, such a demand for fallacies, and, on the other hand, such a supply of them, as for copiousness and variety, taken together, cannot be to be matched elsewhere.

1. In the first place, a thing necessary to the existence of the demand is discussion to a certain degree free.

Where there are no such institutions as a popular assembly taking an efficient part in the Government, and publishing or suffering to be published accounts of its debates,—nor yet any free discussion through the medium of the press,—there is, consequently, no demand for fallacies. Fallacy is fraud, and fraud is useless when every thing may be done by force.

The only case which can enter into comparison with the English Government, is that of the United Anglo-American States.

There, on the side of the *Outs*, the demand for fallacies stands without any difference worth noticing, on a footing similar to that on which it stands under the English Constitution.

But the side of the *Outs* is that side on which the demand for fallacies is by much the least urgent and abundant.

On the side of the *Ins*, the demand for fallacies depends upon the aggregate mass of abuse: its magnitude and urgency depend upon the magnitude of that mass, and its variety upon the variety of the shapes in which abuse has manifested itself.

On crossing the water, fortune gave to British America the relief that policy gave to the fox; of the vermin by which she had been tormented, a part were left behind.

No deaf auditors of the Exchequer: no blind surveyors of melting irons: no non-registering registrars of the Admiralty Court, or of any other judicatory: no tellers, by whom no money is told but that which is received into their own pockets: no judge acting as clerk under himself: no judge pocketing 7000*l.* a-year, for useless work, for which men are forced to address his clerks. No judge, who, in the character of judge over himself, sits in one place to protect, by storms of fallacy and fury, the extortions and oppressions habitually committed in another: no tithe-gatherers exacting immense retribution for minute or never rendered service.

With respect to the whole class of fallacies built upon authority, precedent, wisdom of ancestors, dread of innovation, immutable laws, and many others, occasioned by ancient ignorance and ancient abuses, what readers soever there may be, by whom what is to be found under those several heads has been perused, to them it will readily occur, that in the American Congress the use made of these fallacies is not likely to be so copious as in that August Assembly, which, as the only denomination it can with propriety be called by, has been pleased to give itself that of the Imperial Parliament of Great Britain and Ireland.

CHAPTER IX.

The demand for Political Fallacies :—how created by the state of interests.

In order to have a clear view of the object to which political fallacies will, in the greatest number of instances, be found to be directed, it will be necessary to advert to the state in which, with an exception comparatively inconsiderable, the business of Government ever has been, and still continues to be, in every country upon earth: and for this purpose must here be brought to view a few positions, the proof of which, if they require any, would require too large a quantity of matter for this place: positions which, if not immediately assented to, will at any rate, even by those whom they find most averse, be allowed to possess the highest claim to attention and examination.

1. The end or object in view to which every political measure, whether established or proposed, ought according to the extent of it to be directed, is the greatest happiness of the greatest number of persons interested in it, and that, for the greatest length of time.

2. Unless the United States of North America be virtually an exception, in every known state the happiness of the many has been at the absolute disposal either of the one or of the comparatively few.

3. In every human breast, rare and short-lived ebullitions, the result of some extraordinary strong

stimulus or incitement, excepted, self-regarding interest is predominant over social interest: each person's own individual interest, over the interests of all
other persons taken together.

4. In the few instances, if any, in which, throughout the whole tenour or the general tenour of his life,
a person sacrifices his own individual interest to that
of any other person or persons, such person or persons
will be, a person or persons with whom he is connected by some domestic or other private and narrow
tie of sympathy; not the whole number, or the majority
of the whole number, of the individuals of which the
political community to which he belongs is composed.

5. If in any political community there be any individuals by whom, for a constancy, the interests of all
the other members put together, are preferred to the
interest composed of their own individual interest, and
that of the few persons particularly connected with them,
these public-spirited individuals will be so few, and at
the same time so impossible to distinguish from the
rest, that to every practical purpose they may, without
any practical error, be laid out of the account.

6. In this general predominance of self-regarding
over social interest, when attentively considered, there
will not be found any just subject of regret any more
than of contestation : for it will be found, that but for
this predominance, no such species as that which we
belong to could have existence: and that, supposing it,
if possible, done away, in so much that all persons or
most persons should find respectively, some one or more

persons, whose interest was respectively, through the whole of life, dearer to them, and as such more anxiously and constantly watched over than their own, the whole species would necessarily, within a very short space of time, become extinct.

7. If this be true, it follows, by the unchangeable constitution of human nature, that in every political community, by the hands by which the supreme power over all the other members of the community is shared, the interest of the many over whom the power is exercised, will, on every occasion, in case of competition, be in act or in endeavour sacrified to the particular interest of those by whom the power is exercised.

8. But every arrangement by which the interest of the many is sacrificed to that of the few, may, with unquestionable propriety, if the above position be admitted, and to the extent of the sacrifice, be termed a bad arrangement: indeed, the only sort of bad arrangement: those excepted, by which the interest of both parties is sacrificed.

9. A bad arrangement, considered as already established and in existence, is, or may be termed, *an abuse.*

10. In so far as any competition is seen, or supposed to have place, the interests of the subject many, being on every occasion, as above, in act or in endeavour constantly sacrificed by the ruling few to their own particular interests, hence, with the ruling few, a constant object of study and endeavour, is, the preservation

and extension of the mass of abuse: at any rate such
is the constant propensity.

11. In the mass of abuse, which, because it is so
constantly their interest, it is constantly their endeavour to preserve, is included not only that portion from
which they derive a direct and assignable profit, but
also that portion from which they do not derive any
such profit. For the mischievousness of that from
which they do not derive any such direct and particular profit, cannot be exposed but by facts and observations, which, if pursued, would be found to apply also to that portion from which they do derive
direct and particular profit. Thus it is, that in every
community all men in power, or, in one word, the
Ins, are, by self-regarding interest, constantly engaged
in the maintenance of abuse in every shape in which
they find it established.

12. But whatsoever the *Ins* have in possession, the
Outs have in expectancy. Thus far, therefore, there
is no distinction between the sinister interests of the
Ins and those of the *Outs*, nor, consequently, in the
fallacies by which they respectively employ their endeavours in the support of their respective sinister
interests.

13. Thus far the interests of the Outs coincide with
the interest of the Ins. But there are other points in
which their interests are opposite. For procuring for
themselves the situations and mass of advantages possessed by the Ins, the Outs have one and but one
mode of proceeding. This is the raising their own

place in the scale of political reputation, as compared
with that of the Ins.   For effecting this ascendancy,
they have accordingly two correspondent modes :—the
raising their own, and the depreciating that of their
successful rivals.

14. In addition to that particular and sinister interest which belongs to them in their quality of ruling
members, these rivals have their share in the universal
interest which belongs to them in their quality of
members of the community at large.   In this quality,
they are sometimes occupied in such measures as in
their eyes are necessary for the maintenance of the
universal interest;—for the preservation of that portion
of the universal happiness of which their regard for
their own interests does not seem to require the sacrifice : for the preservation, and also for the increase of
it : for by every increase given to it, they derive advantage to themselves, not only in that character
which is common to them with all the other members
of the community, but, in the shape of reputation, in
that character of ruling members which is peculiar to
themselves.

15. But, in whatsoever shape the Ins derive reputation to themselves, and thus raise themselves to a
higher level in the scale of comparative reputation, it
is the interest of the Outs as such, not only to prevent
them from obtaining this rise, but if possible, and as
far as possible, to cause their reputation to sink.
Hence on the part of the Outs there exists a constant
tendency to oppose all good arrangements proposed

by the Ins. But generally speaking, the better an arrangement really is, the better it will generally be thought to be; and the better it is thought to be, the higher will the reputation of its supporters be raised by it. In so far, therefore, as it is in their power, the better a new arrangement proposed by the Ins is, the stronger is the interest by which the Outs are incited to oppose it. But the more obviously and indisputably good it is when considered in itself, the more incapable it is of being successfully opposed in the way of argument otherwise than by fallacies; and hence in the aggregate mass of political fallacies, may be seen the character and general description of that portion of it which is employed chiefly by the Outs.

16. In respect and to the extent of their share in the universal interest, an arrangement which is beneficial to that interest will be beneficial to themselves: and thus, supposing it successful, the opposition made by them to the arrangement would be prejudicial to themselves. On the supposition, therefore, of the success of such opposition, they would have to consider which in their eyes would be the greater advantage—their share in the advantage of the arrangement, or the advantage promised to them by the rise of their place in the comparative scale of reputation, by the elevation given to themselves, and the depression caused to their adversaries.

But generally speaking, in a Constitution such as the English in its present state, the chances are in a

prodigious degree against the success of any opposition made by the Outs to even the most flagrantly
bad measure of the Ins : much more, of course, to a
really good one.   Hence it is, that when the arrangement is in itself *good*, if with any prospect of success
or advantage, any of the fallacies belonging to their
side can be brought up against the arrangement, and
this without prejudice to their own reputation,—they
have nothing to stand in the way of the attempt.

17. In respect of those *bad* arrangements which
by their sinister interest the Ins stand engaged to
promote, and in the promotion of which the Outs
have, as above, a community of interest,—the part
dictated by their sinister interest is a curious and
delicate one.   By success, they would lessen that
mass of sinister advantage which, being that of their
antagonists in possession, is theirs in expectancy.
They have, therefore, their option to make between
this disadvantage and the advantage attached to a
correspondent advance in the scale of comparative
reputation.   But, their situation securing to them little
less than a certainty of failure, they are, therefore, as
to this matter, pretty well at their ease.   At the same
time, seeing that whatsoever diminution from the
mass of abuse they were to propose in the situation of
Outs, they could not without loss of reputation, unless
for some satisfactory reason, avoid bringing forward,
or at least supporting, in the event of their changing
places with the Ins,—hence it is, that any such defal-

cation which they can in general prevail upon themselves to propose, will in general be either spurious
and fallacious, or at best inadequate :—inadequate,—
and by its inadequacy, and the virtual confession involved in it, giving support and confirmation to every
portion of kindred abuse which it leaves untouched.

CHAPTER X.

Different parts which may be borne in relation to Fallacies.

As in the case of bad money, so in the case of bad arguments, in the sort and degree of currency which they experience, different persons acting so many different parts are distinguishable.

Fabricator, utterer, acceptor : these are the different parts acted in the currency given to a bad shilling : these are the parts acted on the occasion of the currency given to a bad argument.

In the case of a bad argument, he who is *fabricator* must be utterer likewise, or in general it would not make its appearance.   But for one fabricator, who is an utterer, there may be utterers in any number, no one of whom was fabricator.

In the case of the bad argument, as in the case of the bad shilling, in the instance of each actor, the mind is, with reference to the nature and tendency of the transaction, capable of bearing different aspects, which, for purposes of practical importance, it becomes material to distinguish.

1. Evil consciousness, (in the language of Roman lawyers *dolus;* in the language of Roman and thence of English lawyers *mala fides*) : blameable ignorance or inattention, say in one word, 2. *temerity,* (in the same language sometimes *culpa,* sometimes *temeritas*) :

3. Blameless agency, *actus*; which, notwithstanding anymischief that may have been the casual result of it, was free of blame :—by these several denominations are characterized so many habitudes, of which, with relation to any pernicious result, the mind is susceptible.

In the case of the argument, as in the case of the shilling, where the mind is in that state in which the charge of evil-consciousness may with propriety be made, that which the man is conscious of, is, the badness of the article which he has in hand.

In general it is in the case of the *fabricator*, that the mind is least apt to be free from the imputation of evil-consciousness. Be it the bad shilling, be it the bad argument, the making of it will have cost more or less trouble ; which trouble, generally speaking, the fabricator will not have taken but in the design of utterance, and in the expectation of making, by means of such utterance, some advantage. In the instance of the bad shilling, it is certain ; in the instance of the bad argument, it is more or less probable (more probable in the case of the fabricator than in the case of the mere utterer) that the badness of it was known and understood. It is certainly possible that the badness of the argument may never have been perceived by the fabricator, or that the bad argument may have been framed without any intention of applying it to bad purposes. But in general, the more a man is exposed to the action of sinister interest, the more reason there is for charging him with evil-consciousness, supposing him to be aware of the action of the sinister interest.

However the action of the sinister interest may have been either *perceived* or *unperceived*, for without a certain degree of attention a man no more perceives what is passing in his own than what is passing in other minds: the book that lies open before him, though it be the object nearest to him, and though he be ever so much in the habit of reading, may even while two eyes are fixt upon it be read or not read, according as it happens that circumstances have, or have not, called his attention to the contents.

The action of a sinister interest may have been *immediate* or *un-immediate*.

Immediate; it may have been perceived or not perceived: un-immediate; it has, almost to a certainty, been unperceived.

Sinister interest has two *media* through which it usually operates. These are *prejudice* and *authority ;* and hence we have for the immediate progeny of sinister interest, *interest-begotten* prejudice and *authority-begotten* prejudice.

In what case soever a bad argument has owed its fabrication or its utterance to sinister interest, and that interest is not, at the time of fabrication or utterance, perceived, it has for its immediate parent either *in-bred* prejudice or *authority*.

Of the three operations thus intimately connected, viz. *fabrication, utterance,* and *acceptance,* that the two first are capable of having *evil-consciousness* for

their accompaniment is obvious.   As to acceptance,
a distinction must be made before an answer can be
given to the question, whether it is accompanied with
evil-consciousness.

It may be distinguished into *interior* and *exterior*.
Where the opinion, how false soever, is really believed
to be true by the person to whom it has been presented, the acceptance given to it may be termed internal: where, whether by discourse, by deportment,
or other tokens, a belief of its having experienced an
internal acceptance at his hands is, with or without
design on his part, entertained by other persons; in
so far may it be said to have experienced at his hands
an external acceptance.

In the natural state of things, both these modes of
acceptance have place together: upon the *internal*,
the *external* mode follows as a natural consequence.
Either of them is, however, capable of having place
without the other: feeling the force of an argument,
I may appear as if I had not felt it: not having received any impression from it, I may appear as if I
had received an impression of greater or less strength,
whichever best suits my purpose.

It is sufficiently manifest that evil-consciousness
cannot be the accompaniment of internal acceptance;
but it may be an accompaniment, and actually is the
accompaniment of external acceptance, as often as
the external has not for its accompaniment the internal acceptance.

Supposing the argument such that the appellation
of fallacy is justly applicable to it, whatsoever part is

borne in relation to it, viz. fabrication, utterance, or acceptance, may, with propriety, be ascribed to want of probity or want of intelligence.

Hitherto the distinction appears plain and broad enough ; but upon a closer inspection a sort of a mixed, or a middle state between that of evil-consciousness and that of pure temerity—between that of improbity and that of imbecility, may be observed.

This is where the persuasive force of the argument admits of different degrees ; as when an argument which operates with a certain degree of force on the utterer's mind, is in the utterance given to it represented as acting with a degree of force to any amount more considerable.

Thus, a man who considers his opinion as invested only with a certain degree of probability, may speak of it as of a matter of absolute certainty. The persuasion he thus expresses is not absolutely false, but it is exaggerated, and this exaggeration is a species of falsehood.

The more frequent the trumpeter of any fallacy is in its performance, the greater the progress which his mind is apt to make from the state of evil-consciousness to the state of temerity—from the state of improbity to the state of imbecility ; that is, imbecility with respect to the subject matter. It is said of gamblers, that they begin their career as dupes and end as thieves : in the present case, the parties begin with craft and end with delusion.

A phenomenon, the existence of which seems to be not of dispute, is that of a liar by whom a lie of his own invention has so often been told as true, that at length it has come to be accepted as such even by himself.

But if such is the case with regard to a statement composed of words, every one of which finds itself in manifest contradiction to some determinate truth, it may be imagined how much more easily, and consequently how much more frequently, it may come to be the case, in regard to a statement of such nicety and delicacy, as that of the strength of the impression made by this or that instrument of persuasion, of which the persuasive force is susceptible of innumerable degrees, no one of which has ever yet been distinguished from any other, by any externally sensible signs or tokens, in the form of discourse or otherwise.

If substitution of irrelevant arguments to relevant ones is evidence of a bad cause, and of consciousness of the badness of that bad cause, much more is the substitution—of application made to the *will*, to applications made to the *understanding* :—of the matter of punishment or reward, to the matter of argument.

Arguments addressed to the understanding, may, if fallacious, be answered ; and any mischief they had a tendency to produce, be prevented by counter arguments, addressed to the understanding.

Against arguments addressed to the will ; those addressed to the understanding are altogether without effect, and the mischief produced by them is without remedy.

CHAPTER XI.

Uses of the preceding Exposure.

BUT of these disquisitions concerning the state and character of the mind of those by whom these instruments of deception are employed, what, it may be asked, is the practical use?

The use is the opposing such check as it may be in the power of reason to apply to the practice of employing these poisoned weapons. In proportion as the virtue of sincerity is an object of love and veneration, the opposite vice is held in abhorrence:— the more generally and intimately the public in general are satisfied of the insincerity of him by whom the arguments in question are employed, in that same proportion will be the efficiency of the motives by the force of which a man is withheld from employing these arguments.

Suppose the deceptious and pernicious tendency of these arguments, and thence the improbity of him who employs them, in such sort held up to view as to find the minds of men sufficiently sensible of it; and suppose that in the public mind in general, virtue in the form of sincerity is an object of respect, vice in the opposite form an object of aversion and contempt,

the practice of this species of improbity will become
as rare, as is the practice of any other species of improbity, to which the restrictive action of the same
moral power, is in the habit of applying itself with the
same force.

If, on this occasion, the object were to prove the
deceptious nature and inconclusiveness of these arguments, the exposure thus given of the mental character of the persons by whom they are employed,
would not have any just title to be received into the
body of evidence applicable to this purpose.   Be the
improbity of the persons by whom these arguments
are employed ever so glaring, the arguments themselves are exactly what they are, neither better nor
worse.   To employ as a medium of proof for demonstrating the impropriety of the arguments, the improbity of him by whom they are uttered, is an expedient
which stands itself upon the list of *fallacies*, and which
in the foregoing pages has been brought to view.

But on the present occasion, and for the present
purpose, the impropriety as well as the mischievousness of these arguments is supposed to be sufficiently
established on other, and those unexceptionable,
grounds : the object in view now is, to determine by
what means an object so desirable as the general disuse
of these poisonous weapons may in the completest
and most effectual degree be attained.

Now, the mere *utterance* of these base arguments
is not the only, it is not so much as the principal mis

chief in the case. It is the reception of them in the character of conclusive or influential arguments, that constitutes the principal and only ultimate mischief. To the object of making men ashamed to utter them must, therefore, be added, the ulterior object of making men ashamed to receive them : ashamed as often as they are observed to see or hear them,— ashamed to be known to turn towards them any other aspect than that of aversion and contempt.

But if the practice of insincerity be a practice which a man ought to be ashamed of, so is the practice of giving encouragement to—of forbearing to oppose discouragement to that vice : and to this same desirable and useful end does that man most contribute, by whom the immorality of the practice is held up to view in the strongest and clearest colours.

Nor, upon reflection, will the result be found so hopeless as at first sight might be supposed. In the most numerous assembly that ever sat in either house, perhaps, not a single individual could be found, by whom, in the company of a chaste and well-bred female, an obscene word was ever uttered. And if the frown of indignation were as sure to be drawn down upon the offender by an offence against this branch of the law of probity, as by an offence against the law of delicacy, transgression would not be less effectually banished from both those great public theatres, than it is already from the domestic circle.

If, of the fallacies in question, the tendency be

really pernicious, whosoever he be, who by lawful and unexceptionable means of any kind shall have contributed to this effect, will thereby have rendered to his country and to mankind good service.

But whosoever he be, who to the intellectual power, adds the moderate portion of pecuniary power necessary, in his power it lies completely to render this good service.

In any printed report of the debates of the assembly in question, supposing any such instruments of deception discoverable, in each instance in which any such instrument is discoverable, let him, at the bottom of the page, by the help of the usual marks of reference, give intimation of it : describing it, for instance, if it be of the number of those which are included in the present list, by the name by which it stands designated in this list, or by any more apt and clearly designative denomination that can be found for it.

The want of sufficient time for adequate discussion, when carried on orally in a numerous assembly, has in no inconsiderable extent been held out by experience in the character of a real and serious evil. To this evil, the table of fallacies furnishes, to an indefinite extent, a powerful remedy.

There are few men of the class of those who read, to whose memory Goldsmith's delightful novel, the Vicar of Wakefield, is not more or less present. Among the disasters into which the good Vicar is betrayed by his simplicity, is the loss inflicted on him by the craft of Ephraim Jenkins. For insinuating himself into the good opinion and confidence of men of more learning than caution, the instrument he had formed to himself consisted apparently of an extempore sample of recondite learning, in which, in the character of the subject, the cosmogony, and in the character of one of the historians, Sanchoniathon, were the principal figures. On one or two of the occasions on which it was put to use, the success corresponded with the design, and Ephraim remained undetected and triumphant. But at last, as the devil by his cloven foot, so was Ephraim, though in a fresh disguise, betrayed by the cosmogony and Sanchoniathon, to some persons to whose lot it had fallen to receive the same proof of recondite learning, word for word. Immediately the chamber rings, with—" *Your servant, Mr. Ephraim!*"

In the course of time when these imperfect sketches shall have received perfection and polish from some more skilful hand, so shall it be done unto him, (nor is there need of inspiration for the prophecy,) so shall it be done unto him, who in the tabernacle of St. Stephen's or in any other mansion, higher or lower, of similar design and use, shall be so far off his guard as through craft or simplicity to let drop any of these irrelevant, and at one time deceptious arguments: and instead of, Order! Order! a voice shall be heard, followed, if need be, by voices in scores, crying aloud, "Stale! Stale! Fallacy of authority, Fallacy of distrust," &c. &c.

The faculty which detection has of divesting Deception of her power, is attested by the poet,—

" Quære *peregrinum*, vicinia rauca reclamant."

The period of time at which, in the instance of the instruments of deception here in question, this change shall have been acknowledged to have been completely effected, will form an epoch in the history of civilization.

THE END.

Printed by Richard Taylor,
Shoe-Lane, London.

CPSIA information can be obtained
at www.ICGtesting.com
Printed in the USA
LVHW041106140223
739360LV00003B/628